STRUCTURAL, OPTICAL, ELECTRICAL AND MAGNETIC PROPERTIES OF SYNTHESIZED MANGANESE SULFIDE NANOCRYSTALS

Dr. T. VEERAMANIKANDASAMY
Assistant Professor
Department of Electronics and Communication Systems
Sri Krishna Arts and Science College
Coimbatore - 641008, India.

Dr. K. RAJENDRAN
Assistant Professor
Department of Electronics
L.R.G Government Arts College for Women
Tirupur - 641604, India.

Structural, Optical, Electrical and Magnetic Properties of Synthesized Manganese Sulfide Nanocrystals

Copyright © 2019 T Veeramanikandasamy

All rights reserved. No part of this publication may be reproduced, distributed, or transmitted in any form or by any means, including photocopying, recording, or other electronic or mechanical methods, without the prior written permission of the publisher, except in the case of brief quotations embodied in critical reviews and certain other noncommercial uses permitted by copyright law.

ISBN: **9781670377302** (Paperback)

Cover image by T Veeramanikandasamy

First Edition 2019

About the Authors

T Veeramanikandasamy is an Assistant Professor of Electronics and Communication Systems at Sri Krishna Arts and Science College, Coimbatore 641 008. He obtained his Doctorate in Electronics from Bharathiar University, Coimbatore, India. He has 13 years of teaching experience. His current research interests are in Nanomaterials Characterization, Embedded Systems, and Digital Signal Processing. He has published 11 research papers in peer-reviewed international journals. He has presented more than 16 research papers in national/international conferences. He has delivered more than 10 technical lectures in various institutions. He is a Life Member of the Indian Society of Systems for Science and Engineering (ISSE). He received a certificate in Embedded Software Engineer (NSQF-QP ELE/Q1501) from the Electronics Sector Skills Council of India (ESSCI). He has developed various student enrichment courses on Embedded Systems, Digital System Design, Digital Signal Processing, Programmable Logic Controller and IoT with Python.

K. Rajendran is an Assistant Professor of Electronics at LRG Govt. Arts College for Women, Tirupur-641604. He obtained his Doctorate in Electronics from Bharathiar University, Coimbatore. He has produced 13 M.Phil and 6 Ph.D. research scholars and more than 25 research papers to his credit. His area of special interest includes synthesis of Nanomaterials for electronic applications, fabrication of solar cells and Embedded Systems. He has completed one minor research project in the area of solar cells funded by TNSCHE.

PREFACE

Semiconductor nanocrystals are an important class of materials with unique chemical and physical properties owing to the quantum confinement effect. These unique properties of semiconductor nanocrystals have attracted great attention and having a wide optical bandgap which can be engineered through the variation of the material composition and size. Most common nanocrystals are made of Cd and Pb which are toxic. Recent environmental regulations restrict the use of toxic metals and therefore the nontoxic nanocrystalline metals such as Zn, Mn, Cu, and Fe are of great importance. Materials containing manganese are interesting because of their applications in many areas of modern technology. Manganese Sulfide (MnS) is a wide bandgap VIIB–VIA diluted magnetic semiconductor. MnS is a magnetic material having potential use in short-wavelength optoelectronic devices, solar cell applications as a window buffer material and used in anode material for Li-ion batteries. Manganese Sulfide (MnS) was broadly studied because of its magneto-optical properties and wide bandgap nature. Over the years, numerous protocols have been designed for the synthesis of metal sulfide nanocrystals over a range of composition, size, and shapes. But, the synthesis of nanocrystals using wet chemical route is a promising approach receiving increasing attention.

The research presented in this thesis aims to contribute to our understanding of the synthesis and characterization of undoped and Cu-doped MnS nanocrystals. The chapters in this thesis cover the different synthesis strategies in the wet chemical technique used to develop MnS nanocrystals and probe their structural, optical, electrical and magnetic properties as well as some of the possible applications of these materials are addressed. The thesis is divided into seven chapters.

Chapter 1 is designed to provide an introductory overview of semiconductor nanomaterials, magnetic nanostructures, manganese sulfide and its crystal structure, synthesis methods and doping in semiconducting nanocrystals. The chapter aims to place the reader in the scientific context of the thesis.

Chapter 2 describes the brief literature reviews that are reported on the chemical bath deposition technique, RF-sputtering technique, hydrothermal synthetic route, wet chemical technique and SILAR approach available for the preparation of undoped and doped MnS micro/nanomaterials that are established to perform various applications. The motivation, main objectives, and summary of the contributions of present work are also discussed in this chapter.

Chapter 3 describes a brief review of instrumentation, basic theory and experimental setups utilized for various measurements towards the characterization of the synthesized MnS nanocrystals. The characterization of materials is important for understanding their properties and applications. The characterization techniques such as powder X-ray diffraction, Electron microscopy, Fourier transform infrared spectroscopy, UV-Visible spectroscopy, Photoluminescence spectroscopy, Impedance spectroscopy, and Vibrating sample magnetometer are discussed in this chapter.

Chapter 4 describes the effects of refluxing temperature on the preparation of MnS nanocrystals using the wet chemical technique. The structural, optical, electrical and magnetic properties of as-synthesized MnS nanocrystals are presented and compared. XRD results show that the MnS nanocrystals exhibit metastable sphalerite structured β-MnS and metastable wurtzite structured γ-MnS. The average crystallite size and lattice strain are calculated from the XRD data by using the Williamson-Hall equation.

The dislocation density and specific surface area of the samples are also estimated. The morphological changes on the prepared MnS nanocrystals are investigated. UV-Visible optical spectroscopy study is carried out to determine the bandgap energy of as-prepared MnS nanocrystals. The room temperature photoluminescence spectra exhibit two violet-blue emission peaks at around 413nm and 440nm. The electrical measurements on MnS nanocrystals are carried out with the impedance spectroscopy. The dielectric constants, dielectric loss, impedance, electric modulus and AC conductivity with frequencies of the applied field and different temperatures are measured and analyzed its behaviors. The activation energy of electrical processes is also determined at various frequencies. The room temperature Vibrating Sample Magnetometer (VSM) system is used to measure the magnetic properties of materials as a function of the magnetic field. The formed MnS nanocrystals show a paramagnetic behavior. The variations of saturation magnetization (M_s) and coercivity (H_c) of the as-synthesized powders are estimated for the refluxing temperature 55°C, 65°C and 75°C.

Chapter 5 deals with the preparation of MnS nanocrystals by a wet chemical technique employing three different precursor molar ratios (1:1, 1:2 and 2:1) of manganese acetate and thioacetamide. Extensive investigations are carried out on the structural, optical, electrical and magnetic properties of synthesized MnS nanocrystals. The crystallite size, microstrain and dislocation density of synthesized MnS crystals are estimated by modified Scherrer formula and XRD peak broadening analysis. It is also identified the phase transformation from the metastable β-MnS into γ-MnS nanocrystals. The SEM images of MnS nanocrystals are almost spherical. FTIR analysis is performed to study the molecular vibrations and functional groups of the grown MnS nanocrystals. The UV-Visible absorption study reveals to determine the optical bandgap of the MnS nanocrystals and the observed bandgap lies in the range of 4.32 to 4.12 eV.

The two main emission peaks are absorbed in the photoluminescence spectra which are corresponding to the strong blue emissions. The electrical properties of as-prepared samples are investigated using the impedance spectroscopy technique. The permittivity, dielectric loss, complex impedance, electric modulus, AC conductivity, and activation energy are determined at various temperatures as a function of frequency. Maxwell-Wagner type dielectric dispersion is observed from the results of dielectric properties due to interfacial polarization. The complex impedance and modulus analysis confirm that the conduction in these samples is through the grain boundaries. The AC conductivity is a thermally activated process which is used to calculate the activation energy. Analysis of magnetic properties is carried out by using VSM that showed all samples have paramagnetic behavior. The magnetic properties such as saturation magnetization, remanence, coercivity and effective magnetic moment are calculated.

Chapter 6 includes the synthesis and characterization of undoped and Cu-doped MnS nanocrystals prepared by the wet chemical precipitation technique similar to the one used in chapters 4 and 5. The structural, optical, electrical and magnetic properties of these samples are investigated in detail. The crystallite size and lattice strain are calculated using the Williamson-Hall analysis. The change in lattice constants, crystallite size, lattice strain, peak broadening and a small shift in XRD peaks reveal the doping effect of Cu2+ ions in the MnS lattice. The blue shift of the optical bandgap is observed in the Cu-doped MnS nanocrystals compared to the undoped MnS nanocrystals. The photoluminescence spectra on Cu-doping show a quenching of the PL intensity. The frequency and temperature dependence dielectric and impedance properties are carried out using impedance spectroscopy.

The Nyquist plot reveals a semicircle implying the response originated from RC elements which correspond to the grain and grain boundary of the synthesized sample. The AC conductivity increases with increasing frequency which is in correlation with the barrier hopping model. The temperature dependence AC conductivity obeys the Arrhenius rule, where the activation energy calculated from the slope. VSM measurement reveals that the undoped MnS has paramagnetic behavior while the Cu-doped MnS has superparamagnetic behavior. On Cu-doping, the saturation magnetization and remanence increases while the coercivity decreases.

Finally, chapter 7 presents a summary derived from this work.

ACKNOWLEDGEMENTS

I would like to thank my professors and the reviewers of this book,

Dr. K. Sundararaman, CEO, Sri Krishna Institutions, Coimbatore-641 008, India.

Dr. K. Sambath, Head, Department of Electronics and Communication Systems, Sri Krishna Arts and Science College, Coimbatore-641 008, India.

Dr. N. Muthukumarasamy, HOD and Associate Professor, Department of Physics, Coimbatore Institute of Technology, Coimbatore, Tamilnadu.

Dr. M. Manikandan, Assistant Professor, Department of Biotechnology, Sri Krishna Arts and Science College, Coimbatore-641 008, India.

Dr. P. Rameshbabu, Center for Crystal Growth, VIT University, Vellore, Tamil Nadu.

Above all, I thank our family god 'Veerakumarasamy' for giving me all these people to help and encourage me, and for the skills and opportunity to complete this research work.

The author welcomes your suggestions for the improvement of subsequent printings and editions.

- Dr. T. Veeramanikandasamy

CONTENTS

Preface		ii
Acknowledgments		vii
Contents		viii
List of Figures		xii
List of Tables		xvi
Abbreviation Expansion		xvii
List of Notations		xix

Chapter I Introduction
1.1	Nanoscience and nanotechnology	1
1.2	Nanomaterials	4
	1.2.1 Introduction	4
	1.2.2 Semiconductor Nanomaterials	5
	1.2.3 Magnetic Nanostructures	7
1.3	Metal sulfides	8
1.4	Manganese Sulfide Nanocrystals	9
	1.4.1 Introduction	9
	1.4.2 Crystal Structure of MnS	10
1.5	Preparation of MnS Nanostructures	13
	1.5.1 Processing Methods	13
	1.5.2 Process parameters	14
1.6	Doped Semiconductor Nanocrystals	15
	References	16

Chapter II Review of literature
2.1	Introduction	19
2.2	Survey of Previous Work	20
	2.2.1 MnS Thin Films	21
	2.2.2 MnS Powders	24
2.3	Motivation of the Present Work	30
2.4	Objective of the Present Work	32
2.5	Summary of the Contributions	33
	References	35

Chapter III Instrumentation and Characterization Techniques
3.1	Introduction	37
3.2	Structural Characterization of Nanocrystals	38
	3.2.1 X-ray Diffraction (XRD)	38
	3.2.1.1 Bragg's Law	38
	3.2.1.2 X-ray Powder Diffractometer	40
	3.2.1.3 XRD Pattern Analysis	41
	3.2.1.4 Estimation of Crystallite size from XRD	42

		3.2.1.5	Lattice Parameter	44
		3.2.1.6	Specific Surface Area	44
		3.2.1.7	Dislocation Density	45
	3.3.2	Electron microscopy		46
		3.2.2.1	Transmission Electron Microscopy	46
		3.2.2.2	Scanning Electron Microscope (SEM)	49
3.3	Optical Characterization of Nanocrystals			51
	3.3.1	Fourier Transform Infrared Spectroscopy		51
	3.3.2	Ultraviolet-Visible Spectroscopy		53
		3.3.2.1	Beer-Lambert Law	54
	3.3.3	Photoluminescence (PL) Spectroscopy		55
		3.3.3.1	PL Experimental setup	57
3.4	Electrical Characterization of Nanocrystals			58
	3.4.1	LCR Meter		58
	3.4.2	Elementary analysis of Impedance spectra		62
	3.4.3	Dielectric Studies		63
	3.4.4	Electric Modulus		65
	3.4.5	AC Conductivity		66
3.5	Magnetic Characterization of Nanocrystals			68
	3.5.1	Vibrating Sample Magnetometer (VSM)		69
	3.5.2	Magnetic Properties		71
		3.5.2.1	Coercivity	72
		3.5.2.2	Paramagnetism	72
		3.5.2.3	Superparamagnetism	73
	3.5.3	Magnetic anisotropy		75
	3.5.4	Effective magnetic moment		76
3.6	Conclusions			76
	References			78

Chapter IV   **Influence of refluxing temperature on the structural, optical, electrical and magnetic properties of chemically synthesized MnS nanocrystals**

4.1	Introduction			81
4.2	Experimental Details			82
	4.2.1	Materials		82
	4.2.2	Synthesis		83
4.3	Results and Discussion			84
	4.3.1	Structural Properties		84
		4.3.1.1	XRD Analysis	84
		4.3.1.2	SEM Analysis	90
	4.3.2	Optical Properties		92
		4.3.2.1	UV-Visible Absorption Spectra Analysis	92
		4.3.2.2	PL Spectra Analysis	94
	4.3.3	Electrical Properties		96

		4.3.3.1	Dielectric Studies	96
		4.3.3.2	Impedance Studies	100
		4.3.3.3	Electric modulus Studies	105
		4.3.3.4	AC conductivity Studies	108
	4.3.4	Magnetic Properties		111
4.4	Conclusions			114
	References			116

Chapter V — Influence of Mn/S molar ratio on the structural, optical, electrical and magnetic properties of chemically synthesized MnS nanocrystals

5.1	Introduction			119
5.2	Experimental Details			120
	5.2.1	Materials		120
	5.2.2	Synthesis		120
5.3	Results and Discussion			122
	5.3.1	Structural Properties		122
		5.3.1.1	XRD Analysis	122
		5.3.1.2	SEM Analysis	127
		5.3.1.3	FTIR Spectra Analysis	128
	5.3.2	Optical Properties		130
		5.3.2.1	UV-Visible Absorption Spectra Analysis	130
		5.3.2.2	PL Spectra Analysis	132
	5.3.3	Electrical Properties		133
		5.3.3.1	Dielectric Studies	134
		5.3.3.2	Impedance Studies	137
		5.3.3.3	Electric Modulus Studies	140
		5.3.3.4	AC Conductivity Studies	143
	5.3.4	Magnetic Properties		146
5.4	Conclusions			148
	References			151

Chapter VI — Influence of Cu-doping on structural, optical, electrical and magnetic properties of chemically synthesized MnS nanocrystals

6.1	Introduction			153
6.2	Experimental Details			154
	6.2.1	Materials		154
	6.2.2	Synthesis		154
6.3	Results and Discussion			156
	6.3.1	Structural Properties		156
		6.3.1.1	XRD Analysis	156
		6.3.1.2	TEM Analysis	159
		6.3.1.3	FTIR Spectra Analysis	160
	6.3.2	Optical Properties		161
		6.3.2.1	UV-Visible Absorption Spectra	162

			Analysis	
		6.3.2.2	PL Spectra Analysis	163
	6.3.3	Electrical Properties		164
		6.3.3.1	Dielectric Studies	164
		6.3.3.2	Impedance Studies	165
		6.3.3.3	Electric modulus Studies	170
		6.3.3.4	AC conductivity Studies	174
	6.3.4	Magnetic Properties		177
6.4	Conclusions			179
	References			181

Chapter VII **Summary**   183

LIST OF FIGURES

		Page No.
Fig. 1.1:	Several fields in nanotechnology	3
Fig. 1.2:	Classification of nanomaterials based on dimensionality	5
Fig. 1.3:	Crystal structure of MnS	11
Fig. 3.1:	Bragg's Diffraction Scheme	39
Fig. 3.2:	Basic setup of X-Ray Diffractometer	41
Fig. 3.3:	Basic principle of a TEM	48
Fig. 3.4:	Schematic representation of a SEM	50
Fig. 3.5:	Schematic of Fourier transform infrared spectrometer	52
Fig. 3.6:	The electronic energy levels and transitions	53
Fig. 3.7:	Schematic of a dual-beam spectrophotometer	55
Fig. 3.8:	Photoluminescence process in a direct bandgap material	56
Fig. 3.9:	Typical setup for the measurement of PL	57
Fig. 3.10:	AC impedance meter	59
Fig. 3.11:	Basic AC impedance parameters	60
Fig. 3.12:	Schematic diagram of four probe LCR meter	61
Fig. 3.13: (a)-(d)	Complex impedance and admittance spectra for series and parallel combinations of R and C	63
Fig. 3.14:	Electric field interactions with an atom under the classical dielectric model	64
Fig. 3.15:	Moment signals to find out the saddle point of the sample	69
Fig. 3.16:	Schematic of vibrating sample magnetometer	70
Fig. 3.17:	Overview of the size dependence of coercivity exhibited by magnetic particles	71
Fig. 3.18:	Scheme of energy barrier separating states with opposite magnetization due to anisotropy	74
Fig. 4.1:	Various steps involved in the wet chemical processing of MnS nanocrystals	84
Fig 4.2(a):	XRD pattern of MnS nanocrystals synthesized at refluxing temperature 55°C	85
Fig 4.2(b):	XRD pattern of MnS nanocrystals synthesized at refluxing temperature 65°C	86
Fig 4.2(c):	XRD pattern of MnS nanocrystals synthesized at refluxing temperature 75°C	86
Fig. 4.3:	W-H linear plots for the MnS nanocrystals synthesized at 55°C (a), 65°C (b) and 75°C (c) refluxing temperature to determine crystallite size and lattice strain	87
Fig. 4.4:	SEM images of synthesized MnS nanocrystals refluxed at 55°C (a), 65°C (b) and 75°C (c)	91

Fig. 4.5:	Optical absorption spectra of MnS nanocrystals synthesized at different refluxing temperatures (55°C, 65°C and 75°C)		94
Fig. 4.6:	Photoluminescence spectra of MnS nanocrystals synthesized at different refluxing temperatures (55°C, 65°C and 75°C)		95
Fig. 4.7:	Frequency and temperature dependence of dielectric constant for the MnS nanocrystals synthesized at 55°C (a & b), 65°C (c & d) and 75°C (e & f) refluxing temperature		97
Fig. 4.8:	Frequency and Temperature dependence of dielectric loss for the MnS nanocrystals synthesized at 55°C (a & b), 65°C (c & d) and 75°C (e & f) refluxing temperature		99
Fig. 4.9:	Variation of the impedance magnitude and phase angle as function of frequency for the MnS nanocrystals synthesized at 55°C (a & b), 65°C (c & d) and 75°C (e & f) refluxing temperature		101
Fig. 4.10:	Equivalent circuit model of as-prepared MnS nanocrystals from impedance analysis		102
Fig. 4.11:	Cole-Cole plots for the MnS nanocrystals synthesized at 55°C (a), 65°C (b) and 75°C (c) refluxing temperature		104
Fig. 4.12:	Variation of M' and M'' with frequency at different temperatures for the MnS nanocrystals synthesized at 55°C (a & b), 65°C (c & d) and 75°C (e & f) refluxing temperature		106
Fig. 4.13:	Relaxation time (τ) of as-prepared MnS nanocrystals with 55°C, 65°C and 75°C refluxing temperature		107
Fig. 4.14:	Frequency dependence of the AC conductivity (σ_{ac}) for the MnS nanocrystals synthesized at 55°C (a), 65°C (b) and 75°C (c) refluxing temperature		109
Fig. 4.15:	Variation of σ_{ac} against the $10^3/T$ of MnS nanocrystals synthesized at 55°C (a), 65°C (b) and 75°C (c) refluxing temperature		110
Fig. 4.16:	Variation of activation energy with frequency for the MnS nanocrystals synthesized at 55°C, 65°C and 75°C refluxing temperature		111
Fig. 4.17:	Magnetic moment vs. magnetic field plots of the MnS nanocrystals synthesized at 55°C, 65°C and 75°C refluxing temperature		112
Fig. 5.1:	A schematic diagram of the formation of MnS nanocrystals		121
Fig. 5.2:	XRD patterns of MnS nanocrystals with different precursor molar ratios (a) 1:1, (b) 1:2 and (c) 2:1		124

Fig. 5.3:	Linear plots of modified Scherrer equation and obtained intercepts for the MnS nanocrystals with different precursor molar ratios (a) 1:1, (b) 1:2 and (c) 2:1	125
Fig. 5.4:	SEM images of the sample synthesized with different precursor molar ratios (a) 1:1, (b) 1:2 and (c) 2:1	128
Fig. 5.5:	FTIR spectra of the MnS nanocrystals with different precursor molar ratios (a) 1:1, (b) 1:2 and (c) 2:1	129
Fig. 5.6:	UV-Visible absorption spectra of the MnS nanocrystals with 1:1, 1:2 and 2:1 molar ratio	131
Fig. 5.7:	PL spectra of the MnS nanocrystals with 1:1, 1:2 and 2:1 molar ratio	133
Fig. 5.8:	Frequency and temperature dependence of dielectric constant for the MnS nanocrystals with different precursor molar ratios (a & b) 1:1, (c & d) 1:2 and (e & f) 2:1	135
Fig. 5.9:	Frequency and temperature dependence of dielectric loss for the MnS nanocrystals with different precursor molar ratios (a & b) 1:1, (c & d) 1:2 and (e & f) 2:1	136
Fig. 5.10:	Frequency dependence of |Z| at different temperatures for the MnS nanocrystals with different precursor molar ratios (a) 1:1, (b) 1:2 and (c) 2:1	138
Fig. 5.11:	Complex impedance plots of the MnS nanocrystals with different precursor molar ratios (a) 1:1, (b) 1:2 and (c) 2:1	139
Fig. 5.12:	Equivalent circuit of the as-prepared samples from the impedance spectra	140
Fig. 5.13:	Frequency dependence of M' and M'' at different temperature for the MnS nanocrystals with different precursor molar ratios (a & b) 1:1, (c & d) 1:2 and (e & f) 2:1	142
Fig. 5.14:	Frequency and temperature dependence of AC conductivity for the MnS nanocrystals with different precursor molar ratios (a & b) 1:1, (c & d) 1:2 and (e & f) 2:1	144
Fig. 5.15:	Variations in activation energy with frequency for the as-prepared MnS nanocrystals	145
Fig. 5.16:	Magnetic moment vs. magnetic field plots of the MnS nanocrystals with 1:1, 1:2 and 2:1 molar ratio	146
Fig. 6.1:	Various steps involved in the experimental procedures for undoped and Cu-doped MnS nanocrystals	155
Fig. 6.2:	XRD patterns of (a) undoped MnS and (b) Cu-doped MnS nanocrystals	157
Fig. 6.3:	W-H plots for (a) undoped MnS and (b) Cu-doped MnS nanocrystals	158

Fig. 6.4:	TEM images of (a) undoped MnS and (b) Cu-doped MnS nanocrystals	159
Fig. 6.5:	FTIR spectra of (a) undoped MnS and (b) Cu-doped MnS nanocrystals	161
Fig. 6.6:	Optical properties of MnS and Cu-doped MnS nanocrystals (a) UV-Visible absorbance spectra (b) Photoluminescence (PL) spectra	163
Fig. 6.7:	Variation of dielectric constant with frequency at different temperatures for (a) undoped MnS and (b) Cu-doped MnS nanocrystals	166
Fig. 6.8:	Variation of dielectric loss with frequency at different temperatures for (a) undoped MnS and (b) Cu-doped MnS nanocrystals	167
Fig. 6.9:	Variation of impedance |Z| with frequency at different temperatures for (a) undoped MnS and (b) Cu-doped MnS nanocrystals	169
Fig. 6.10:	Complex impedance plots for (a) undoped MnS and (b) Cu-doped MnS nanocrystals	170
Fig. 6.11:	Relaxation time with temperature plot for the undoped MnS and Cu-doped MnS nanocrystals	171
Fig. 6.12:	Frequency dependence of M' at various temperatures for (a) undoped MnS and (b) Cu-doped MnS nanocrystals	172
Fig. 6.13:	Frequency dependence of M" at various temperatures for (a) undoped MnS and (b) Cu-doped MnS nanocrystals	173
Fig. 6.14:	Variation of AC conductivity with frequency at different temperatures for (a) undoped MnS and (b) Cu-doped MnS nanocrystals	175
Fig. 6.15:	Inverse temperature dependence of AC conductivity at various frequencies for (a) undoped MnS and (b) Cu-doped MnS nanocrystals	176
Fig. 6.16:	Magnetization curve as a function of applied magnetic field at RT for the undoped MnS and Cu-doped MnS nanocrystals	177

LIST OF TABLES

		Page No.
Table 4.1:	XRD data analysis for Crystallite size, Micro strain and Dislocation density of MnS nanocrystals synthesized at different refluxing temperatures	89
Table 4.2:	Optical band gap of MnS nanocrystals synthesized at different refluxing temperatures	94
Table 4.3:	Equivalent circuit parameters of impedance spectra for the MnS nanocrystals synthesized at different refluxing temperatures	105
Table 5.1:	XRD data analysis for Crystallite size, Micro strain and Dislocation density of the MnS nanocrystals with 1:1, 1:2 and 2:1 molar ratio	124
Table 5.2:	Values of the equivalent circuit parameters deduced from the impedance spectra for the MnS nanocrystals with 1:1, 1:2 and 2:1 molar ratio	141
Table 5.3:	Magnetic properties of the MnS nanocrystals with 1:1, 1:2 and 2:1 molar ratio	147
Table 6.1:	Values of the equivalent circuit parameters deduced from the impedance spectra for the undoped MnS and Cu-doped MnS nanocrystals	169
Table 6.2:	M_s, M_r, M_c and Magnetic moment of undoped MnS and Cu-doped MnS nanocrystals	178

ABBREVIATION EXPANSION

0D	0 dimensional
1D	1 dimensional
2D	2 dimensional
3D	3 dimensional
A	Absorbance
AC	Alternating current
AFM	Atomic Force Microscope
B	Magnetic Induction (Flux)
BNC	Bayonet Neill-concelman Connector
BS	Saturation Flux Density
BSE	Backscattered electrons
CB	Conduction Band
CBD	Chemical Bath Deposition
Cp	Parallel capacitance
CRT	Cathode Ray Tube
Cs	Series capacitance
CuS	Cupper sulfides
CVD	Chemical Vapour Deposition
D	Dissipation factor
DC	Direct current
DDTC	Diethyldithiocarbamate
DMS	Diluted Magnetic Semiconductors
EDAX	Energy Dispersive X-ray Spectroscopy
FEG	Field Emission Gun
EPR	Electron paramagnetic resonance
F	Magnetic Force
FT-IR	Fourier Transforms Infrared Spectroscopy
FWHM	Full Width at Half Maximum
GMR	Giant magnetoresistance
H	Magnetic Field
Hc	Coercive field strength
HCUR	High current terminal
JCPDS	Joint committee on powder diffraction standards
K	Kelvin
LCR	Inductance Capacitance Resistance
LCUR	Low current terminal
LED	Light Emitting Diode
Ln	Natural logarithm
LUMO	Lowest Unoccupied Molecular Orbital
m	Magnetic Dipole Moment
M	Molar concentration/ Magnetization/Modulus
MBE	Molecular beam epitaxy
MD	Multi domain
MNPs	Magnetic nanoparticles
MnS	Manganese Sulfide
MnSe	Manganese Selenide
Mr	Remanence Point
MRI	Magnetic Resonance Imaging
Ms	Magnetic Saturation

MS	Metal sulfides
MSe	Metal selenides
NCs	Nanocrystals
NiS	Nickel sulfides
NLLS	Nonlinear least square fitting method
NMR	Nuclear Magnetic Resonance
NP	Nanoparticle
PAM	Porous alumina membrane
PL	Photoluminescence
PM	Paramagnetic
QD	Quantum Dot
QW	Quantum Well
R	Resistance
RF	Radio Frequency
rpm	Revolutions Per Minute
Rs	Series resistance
RS	Rock salt
RT	Room Temperature
SD	Single domain
SEM	Scanning Electron Microscopic
SP	Superparamagnetic
T	Transmittance/Temperature
TB	Blocking temperature
TC	Curie temperature
TEA	Triethanolamine
TEM	Transmission Electron Microscope
UDM	Uniform Deformation Model
UV	Ultraviolet
UV-VIS	Ultraviolet-visible spectroscopy
V	Particle Volume
VB	Valance Band
VSM	Vibrating Sample Magnetometer
W	Wurtzite
WH	Williamson Hall
Xm	Magnetic Susceptibility
XRD	X-ray Diffraction
Y	Admittance
Z	Impedance
ZB	Zinc blend

LIST OF NOTATIONS

α	Absorption coefficient
σ_{ac}	AC conductivity
E_a	Activation energy
A^o	Angstrom
k_B	Boltzmann constant
θ	Bragg's angle
R_{Bragg}	Bragg factor
σ	Conductivity
T_c	Curie temperature
σ_{dc}	DC conductivity
tanδ	Dielectric loss tangent
emu	Electromagnetic unit
emu/g	Electromagnetic unit/gram
eV	Electron volt
f	Frequency
R_g	Grain resistance
R_{gb}	Grain boundary resistance
Hz	Hertz
h	Hour
Z″	Imaginary part of impedance
M″	Imaginary part of Modulus
kHz	Kilo Hertz
kOe	Kilo Oersted
a, b, c	Lattice parameters
ω_{max}	Maximum relaxation frequency
m	Meter
μm	Micrometer
μs	Micro second
ms	Milli second
min	Minute
$molL^{-1}$	Mole/liter
nm	nanometer
Oe	Oersted
Ω	Ohms
μ_r	Permeability
μ_0	Permeability of Free Space
Φ	Phase angle
ν	Photon frequency
h	Planck constant
Z′	Real part of impedance
M′	Real part of modulus
ε_r	Relative dielectric constant
ω	Relaxation frequency

Chapter I
INTRODUCTION

1.1 Nanoscience and Nanotechnology

Nanoscience is an emerging area of science. It is the study of phenomena and manipulations of materials at atomic and molecular scales, where properties differ significantly from a large scale [1]. At the nanoscale, physics, chemistry, biology, material science and engineering converge toward the same principles and tools. As the size of matter is reduced to tens of nanometers, the two main reasons for change in behavior are an increased relative surface area and the dominance of quantum effects. An increase in surface area (per unit mass) will result in a consequent increase in chemical reactivity. The quantum effects begin to play a main role in change of material's optical, magnetic and electrical properties. The other effects such as surface tension or stickiness are important, which are also affect physical and chemical properties. Nanoscience is concerned with understanding these effects and their influences on the properties of material [2]. In 1959, the promise of nanotechnology was sketched out by Nobel Prize laureate Richard Feynman in his famous talk, "There's Plenty of Room at the Bottom".

Since then, the concepts of molecular nanotechnology have extended to such as "molecular engineering" by Eric K. Drexler and "molecular electronics" by Mark A. Ratner, etc. Recently, the area of nanotechnology has rapidly developed because enormous possibilities have opened to manipulate the molecular synthesis and movement [3].

Nanotechnology controls the structure of matter at the nanoscale to produce new materials and devices with unique properties. However, some of these technologies have limited control over structure at the nanoscale, but these are being used to produce valuable products. The nanomaterials are also being further developed to produce more complicated products with structure in controlled manners. Nanotechnology is very discrete field, ranging from extensions of conventional device physics to completely new approaches based upon molecular self-assembly [4].

Nanotechnology involves application of numerous scientific fields comprises biomedical sciences, surface science, semiconductor physics, optics, electronics, magnetism, energy storage and electrochemistry [5]. A various range of applications of nanotechnology are shown in Fig. 1.1. The US National Nanotechnology Initiative (NNI) defines a technology as nanotechnology only if it involves all of the following [6]:

- Research and technology development involving at least one dimension structures in approximately the 1-100 nm range and frequently with atomic precision.
- Design of devices and systems using structures that have novel properties and functions because of nanometer scale dimensions.
- Ability to control or manipulate on the atomic scale.

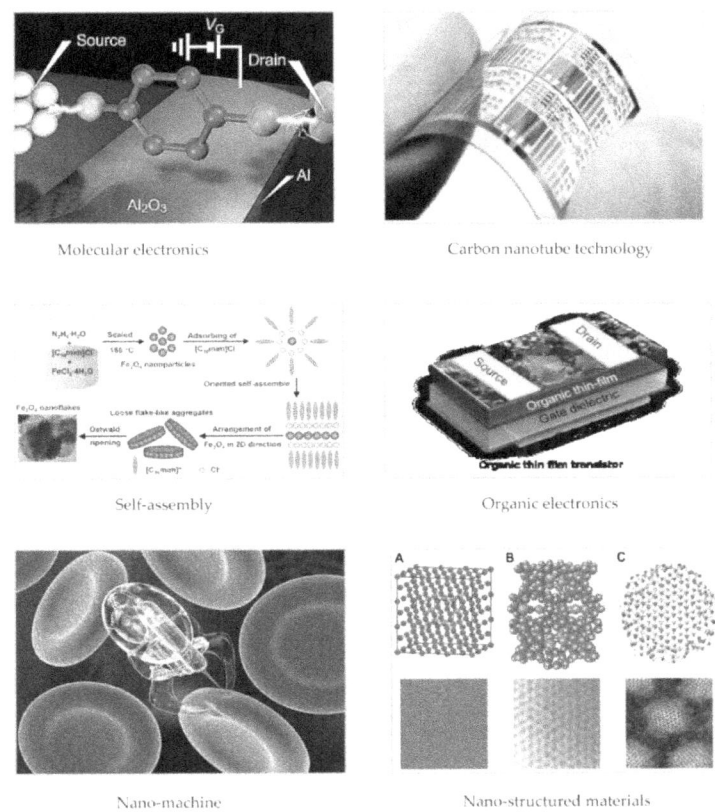

Fig. 1.1: **Several fields in nanotechnology**

There are many ways used to create nanostructured materials, which are usually divided into two main strategies, top-down and bottom-up approaches [7]. The traditional top-down approach is start from bulk materials, which are progressively miniaturized up to the desired smaller size. Alternatively, in bottom-up approach, nanostructured materials are created from building blocks of atoms, molecules, clusters or nanoparticles in a controlled manner, governed by chemical methods.

Therefore, the idea of creating artificial substances and materials with unique features by using the bottom-up approach is increasingly encouraged for the development of new and multi-functional materials.

1.2 Nanomaterials

1.2.1 Introduction

A study on nanomaterials is the engineering of materials with enhanced properties through the controlled synthesis and assembly of material at nanoscale level. The nanostructured materials are in the form of thin films, coatings, powders and as fillers for composite materials. The ability to control the nanostructure of the materials can result in enhanced properties in macroscopic levels like increased hardness, ductility, selective absorption showing more efficient optical, magnetic and electronic behavior.

As shown in Fig. 1.2, Nanomaterials can be classified as (1) zero-dimensional (0-D), (2) one-dimensional (1-D), (3) two-dimensional (2-D), and (4) three-dimensional (3-D). The 0-D nanomaterials are materials wherein all the dimensions are confined to the nanoscale, *i.e.*, no dimensions are larger than 100 nm. They are commonly named as nanoparticles or quantum dots if the particles are sufficiently small and quantum effects are observed. 1-D nanomaterials, however, have two dimensions at the nanoscale and no confinement along another dimension of the materials.

They include a number of types (*e.g.* nanowires, nanopores/nanotubes, nanorods, nanofibres and etc) according to their morphologies. 2-D nanomaterials are materials in which two of the dimensions are not confined to the nanoscale. Therefore, they often exhibit plate-like morphologies, such as nanofilms, nanosheets and nanodisks. 3-D nanomaterials are also known as bulk nanomaterials [8]. They are classified as nanomaterials because they possess nanocrystalline structure although they are not confined to the nanoscale in any dimension.

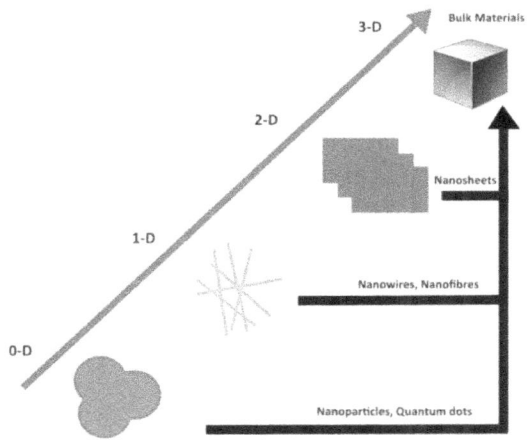

Fig. 1.2: Classification of nanomaterials based on dimensionality

1.2.2 Semiconductor Nanomaterials

During the past decade, tremendous advances in colloid chemistry have led to the preparation of high quality nanometer sized semiconductor materials. A semiconductor is a material which has electrical conductivity intermediate between conductors and insulators. There are a number of ways to classify semiconductor materials.

Most semiconductors are inorganic crystalline solids; however, organic semiconductors have attracted attention in recent years. According to composition, semiconductor materials can be classified into elemental and compound semiconductors. Elemental semiconductors are semiconductors that are made out of a single element, such as silicon and germanium. Compound semiconductors are made out of two or more elements [9]; examples include ZNO, ZNS, CdS, GaN, MnS, etc. Compound semiconductors can be further classified as binary (*e.g.* ZnO, TiO_2, Cu_2S, MnS, MnO, etc.), ternary (*e.g.* $CuInS_2$, CdZnTe), quaternary (*e.g.* Cu_2ZnSnS_4) and quinary [10].

Another popular semiconductor material classification is intrinsic or extrinsic semiconductors. Intrinsic semiconductors are chemically pure and other hand, extrinsic semiconductors have impurities. The addition of other atoms is called 'doping', which can alter the electronic behavior of the material. Depending on the type of doping, extrinsic semiconductors are further classified as n-type and p-type. In addition to the above classifications, semiconductors can be divided into group IV semiconductors, III-V semiconductors, II-VI semiconductors, I-VII semiconductors, and so on, according to respective groups in the periodic table [11].

Among a number of methods for the preparation of II-VII semiconductors, colloidal wet chemical method offers an inexpensive and simple means to synthesize such particles with good control over size and size distribution by optimizing various parameters [12]. Many researchers use ammonia for its dual role of forming complex metal ion and varying the pH of the reaction bath.

This thesis is primarily concerned with the investigation of the structural, optical, electrical and magnetic properties of manganese sulfide semiconductor nanocrystals synthesized from wet chemical preparation method. Due to the unique optical and electrical properties, nanocrystals may play a key role in the emerging new fields of nanotechnology. This area of research was founded in the early 1980s when researchers at Bell Laboratories [13] in the United States and the Yoffe Institute [14, 15] in Russia, then the U.S.S.R, independently described the properties of nanometer sized quantum confined semiconductor crystals. This work was quickly followed by studies on colloidal samples [16-18], leading to a further understanding of their optical and physical properties.

The semiconductor nanocrystals playing a vital role in the semiconductor physics as they show wide range of applications in light emitting diodes [19-21], optoelectronic devices [22, 23], solar energy conversions, single-electron transistors [24, 25] and fluorescent tags for biological imaging applications [26, 27].

1.2.3 Magnetic Nanostructures

A magnetic nanoparticle has a net moment that is the sum of the spins of all its constituent atoms. These atoms are organized within a crystalline structure that determines their magnetic properties. Magnetic nanostructure show a wide variety of unusual magnetic properties compared to the respective bulk materials, which arise from finite size and surface effects. They reveal unique phenomena such as such as high field irreversibility, high saturation field, superparamagnetism, extra anisotropy contributions and etc.

Inside magnetic material, the spins form domains where the individual moments of the atoms are aligned which each other, and their formation is highly related with the size of the particle. When the size of the particle decreases the number of domains also decreases and it becomes a single domain. If the particle size is reduced, there is a critical volume below which it costs more energy to create a domain wall than to support the external magnetostatic energy of the single domain state.

Recently, the synthesis of nanostructured magnetic materials has been an area of intense study, due to new mesoscopic properties shown by quantum-sized particles located in the transition region between atoms and bulk solids. Quantum size effects and the large surface area of magnetic nanoparticles considerably change some of the magnetic properties and quantum tunneling of magnetization [28]. Several research groups are engaged in investigations of the magnetic metal sulfide and oxide nanoparticles because of their technological applications in magnetic and microwave devices, magnetic recording media and etc.

A number of magnetic devices are currently in commercial use. The magnetic storage media used in today's commercial hard disks consist of homogeneous polycrystalline magnetic films. In biomedicine, magnetic nanoparticles (MNPs) serve as contrast enhancement agents for Magnetic Resonance Imaging (MRI) [29], selective probes for bio-molecular interactions and cell sorters.

Nanoparticles of magnetic materials are also find applications as nucleators, catalyst for the growth of high-aspect-ratio nanomaterials, and toxic waste remediation [30]. In magnetic nanoparticles some of the new phenomenon like spin canting and superparamagnetism can be realized which may not be seen in bulk magnetic particles. The saturation magnetization of the magnetic nanoparticles is found to have less value when compared to bulk magnetic particles. However, the coercivity of the magnetic nanoparticles is more value than that of bulk magnetic particles [31].

1.3 Metal Sulfides

Almost, all known elements react with sulfur, forming sulfides and poly sulfides. It has been attempted to discuss at least the most key features of some binary sulfides. Since 70's and 80's the synthesis and characterization of new binary, ternary and quaternary type of metal sulfides have received significant attention. The important technological applications found for many of these materials as well as their remarkable relationships in structure and properties. These applications force the effort in sophisticated synthesis and characterization of sulfide compounds.

All metal can be obtained as metal sulfide or metal oxide compounds. In addition to resemblance between metal sulfides and corresponding metal oxides, the structure and bonding of most binary metal sulfides differ significantly from those of corresponding metal oxides.

The difference, primarily reside in the higher covalence of metal sulfur interactions relative to metal oxygen. The lower electro-negativity of the sulfur relative to the oxygen leaves the valence 3s and 3p sulfur bands much closer in energy to the transition metal d-orbital manifold [32] (i.e., greater covalence). Transition metal binary sulfides have wide application area. Zinc sulfides (ZnS) are used as photocatalyst, nanocables, electroluminescent or phosphorescent materials, fluorescent labeling materials, in solar batteries, cathode ray luminescent, in laser diode, in optical recording, in blue LED's and etc [33-35]. Cupper sulfides (CuS) are used in solar cells and in device for rectifying alternating current [36-37]. Nickel sulfides (NiS) are used as transformation-toughening agent for materials used in semiconductor applications [38], photo electrochemical solar cells, IR detectors, sensors, as agents for air and water purification, as a cathode material in rechargeable lithium batteries [39]. Cobalt sulfides (CoS) are used as semiconductor photo electrode arrays in unassisted photolytic water splitting equipment [40].

1.4 Manganese Sulfide Nanocrystals
1.4.1 Introduction

The manganese sulfide (MnS) crystalline semiconductors have attracted the attention of many researchers due to their potential applications in optoelectronic [41, 42], optical mass memories [43-45], antireflection coatings, spintronics, solar cell applications, blue green light emitter applications [46] and is a promising candidate for many technological applications due to its wider band gap, photoconductive and magnetic behavior. This material is also having interesting photoluminescence properties and hence finds large number of applications in optoelectronic devices.

The MnS belongs to VIIB–VIA diluted magnetic p-type semiconductors [47]. It can be used as an effective nontoxic substitute for cadmium sulfide in solar cell applications. This material not only eliminates toxic cadmium but also improve light transmission in the blue wavelength region on having band gap wider than that of CdS. Manganese Sulphide (MnS) is one of the promising Diluted Magnetic Semiconductors (DMS) materials and has been extensively studied because of its unique magneto-optical properties [48]. Mn is a transition metal with the atomic number 25, with electron configuration [Ar] $4s^2 3d^5$. It can exhibit a number of oxidation states +2, +3, +4, +6 and +7. Mn has five unpaired electrons in the d-shell, making it an excellent candidate as a dopant and its magnetic properties can be incorporated into some metal oxides and sulfides to produce DMS.

1.4.2 Crystal Structure of MnS

As an important magnetic semiconductor material, Manganese chalcogenides (MnX, X = S, Se, Te) exhibit a range of important magneto-optical properties that result from their crystal structures [49]. These are paramagnetic compounds having five unpaired spins; owning to strong electron correlation, they do not form energy bands and become antiferromagnetic. The monosulfide MnS, the disulfide MnS_2, and possibly the trisulfide MnS_3, exist. The structures and phase relations of the MnS system are similar to the zinc and cadmium sulfide at elevated pressures and this is probably a consequence of the spherical symmetry of the high spin d^5 Mn^{2+} ion which is present in MnS and MnS_2. Hence the charge at Mn^{2+} sites is reduced, and metal-metal bonds are increasingly favored [50].

MnS has the NaCl structure. Manganese (II) sulfide and MnS can crystallize into three kinds of structural forms: The meta-stable beta (β) and gamma (γ) phases are pink form at low temperatures with tetrahedrally coordinated zinc blende or sphelerite (SG:F43m - fcc) and wurtzite (SG:P6,mc - hexagonal) crystal structures respectively [51, 52]. The thermodynamically stable alpha (α) phase forms at a relatively higher temperature with octahedrally coordinated rock salt structure (SG:Fm3m) with green color.

Fig. 1.3 shows the crystal structure of Manganese sulfide (MnS). MnS is a magnetic, direct ultra wide band gap magnetic semiconductor with a band gap of Eg= 3.2 eV and a temperature coefficient dEg/dT = -2 meV/K. In all forms, each Mn^{2+} ion has 12 Mn^{2+} nearest neighbors, but different numbers of anion neighbors (6 for α-MnS, 4 for β-MnS). In α-MnS, nearest neighbor Mn^{2+} ions are bonded by a 90° linkage through a S^{2-} ion. The next nearest neighbors are bonded by 180° Mn-S-Mn linkages. In the β forms, nearest neighbor Mn^{2+} ions are bonded tetrahedral [53].

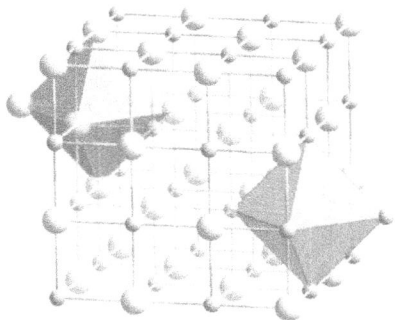

a) α-MnS : Rock salt structure

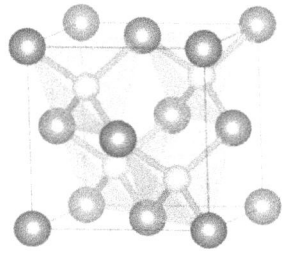

b) β-MnS: Sphalerite structure

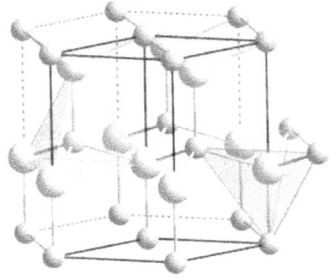

c) γ-MnS: Wurtzite structure

Fig. 1.3: Crystal structure of MnS

In sphelerite structure, the manganese sulfide crystals are composed of equal number of Mn^{2+} and S^{2-} ions [54]. The radii of the two ions (Mn^{+2} = 91 pm and S^{-2} = 184 pm) led to the radius (r+ / r–) as 0.49 which suggests a tetrahedral arrangement. The manganese ions are arranged in ccp arrangement. Manganese ions occupy tetrahedral hole. Only half of the tetrahedral holes are occupied by Mn^{2+} so that the formula of the manganese sulfide is MnS. Since the void is tetrahedral, each manganese ion is surrounded by four sulfide ions and each sulfide ion is surrounded tetrahedrally by four manganese ions. Thus manganese sulfide has 4:4 co-ordinations. There are four Mn^{2+} ions and four S^{2-} ions per unit cell as below:

No. of S^{2-} ions = 8 (at corners) × 1/8 + 6 (at face centres) × 1/2 = 4.
No. of Mn^{2+} ions = 4 (within the body) × 1 = 4.

In wurtzite structure, the sulfide ions have HCP arrangement and manganese ions occupy tetrahedral voids [55]. Only half the alternate tetrahedral voids are occupied by Mn^{2+}. Coordinate number of Mn^{2+} ions as well as S^{2-} ions are 4. Thus, this structure has 4: 4 co-ordinations.

No. of Mn^{2+} ions per unit cell = 4 (within the unit cell) × 1 + 6 (at edge centres) × 1/3 = 6.
No. of S^{2-} ions per unit cell = 12 (at corners) × 1/6 + 2 (at face centres) × 1/2 + 3 (within the unit cell) × 1 = 6.
Thus, there are 6 formula units per unit cell.

In rock salt structure, sulfide ions are arranged in cubic close packing (ccp). In this arrangement, S^{2-} ions are present at the corners and at the centre of each face of the cube (fcc) [56]. The manganese ions are present in all the octahedral holes. The number of octahedral holes in ccp structure is equal to the number of anions, every octahedral hole is occupied by Mn^{2+} ions. There are six octahedral holes around each sulfide ions, each S^{2-} ion is surrounded by 6 Mn^{2+} ions. Similarly each Mn^{2+} ion is surrounded by 6 S^{2-} ions. Therefore, the coordination number of S^{2-} as well as of Mn^{2+} ions is six. This is called 6:6 co-ordinations.

No of Manganese ions = 12 (at edge centres) × 1/4 + 1 (at body centre) × 1= 4.

No of Sulfide ions = 8 (at corner) × 1/8 + 6 (at face centres) × 1/2 = 4.

Thus, the number of MnS units per unit cell is 4.

The electronic spectra of three forms of MnS show that there is considerable mixing of the manganese 3d orbitals with sulfide orbitals. The α-MnS shows an antiferromagnetic phase transition temperature at T_N = 130K due to the correlations between the Mn^{2+} spins, and a paramagnetic moment $p$eff = 5.6μB [57]. In the paramagnetic phase, the MnS is a p-type semiconductor with an activation energy E = 0.3 eV [58].

1.5 Preparation of MnS Nanostructures
1.5.1 Processing Methods

The various methodologies used to produce MnS nanostructure are discussed in this section. It provides the necessary background information with respect to MnS powder production techniques, used to date. It is important to note that each method is greatly dependent on their raw materials.

The stoichiometry (optimum concentration) of the MnS precipitate and its purity depends on the quality of the raw materials used. The production of MnS powders may be classified by the following main headings:

- Wet Chemical Synthesis (such as, Precipitation, Solvothermal, Hydrothermal, CBD and Sol-Gel techniques)
- Dry Chemical Synthesis (such as, solid-state reactions and mechanochemical synthesis)
- Vapour Phase Reactions (such as, spray and freeze-drying)

Wet methods of producing nanostructured materials win the increasing confidence of material scientists and technologists.

Wet chemistry approaches allow the control of the particle growth and pore structure parameters of materials up to several nanometers. A simple adaptation of wet methods to the fabrication conditions makes them promising for a wide range of materials in spintronics, solar cells, implants, medical equipments, catalysts and fine grain ceramics. Since the present research utilizes the wet chemical precipitation method to synthesize phase pure MnS, this method is discussed in this work. The main advantage associated to the wet process, is that by-product consists primarily of water and the probability of contamination during processing is very low. Low processing costs has also been reported [59]. Its disadvantage is that the resulting product can be greatly affected by even a slight difference in the reaction conditions.

1.5.2   Process Parameters

In wet-chemistry approach, it is essential to carefully control the process parameters: stirring rate, process temperature, concentration of components and their quantitative ratio, density, pH, viscosity and other parameters which describe the state of the liquid mediums.

This technique could be considered as soft chemistry because they do not need high temperature and pressure, they do not need to use a large number of expensive energy carrier for their technological realization and can be attributed to eco-technology [60]. In this work for preparing MnS nanocrystals, the process parameters associated with the wet chemical method are reaction temperature and precursor molar concentration; in relation to the effect they impose upon the final MnS nanocrystals obtained. These process variables are controlled, in order to develop MnS nanocrystals of optimum quality.

The ZB-MnS and WZ-MnS NCs have fabricated by wet chemical approach. The influences of size and crystal structure to the optical, electrical and magnetic properties of MnS NCs have been reported.

It was found that the crystal structure and its properties of MnS NCs could be controlled by simply varying the reaction temperature and the molar ratio of precursors. Meanwhile, the corresponding morphologies of MnS NCs also have experienced, which indicate a close relationship between the crystal structure and growth behavior of nanocrystalline.

1.6 Doped Semiconductor Nanocrystals

Impurities or dopants can strongly modify the electronic, optical and magnetic properties of semiconductors. Doped semiconductor nanocrystals not only retain nearly all advantages of intrinsic nanocrystals, but also eliminate their self-quenching due to reabsorption/energy transfer. The energy from absorbed photons can be efficiently transferred to impurities, suppressing unwanted reactions on the nanocrystal surface or reabsorption by the band gap because of the small stokes shift between absorption and emission wavelengths. The energy transfer to impurities greatly improves their thermal and chemical stability. In addition, Doped nanocrystals can be made from less harmful elements than those currently used (Class A elements Cd, Hg, and Pb) [61]. The emission wavelength of doped semiconductor nanocrystals can also be tuned [62]. The recent investigations describes a process for the preparation of transition metal ion (various transition metal ions such as Mn^{2+}, Co^{2+}, Cu^{2+} and Ni^{2+}) doped semiconductor nanocrystals, where the reactions take place at a temperature less than 200°C. Using wet chemical synthesis, the effect of Cu-doping with MnS nanocrystals have been investigated in this work. The said doped nanocrystals are stable under UV radiation in both solution and precipitated solid form.

References

1. Shivani Soni, Amandeep Salhotra, Mrutyunjay Suar, IGI Global, 2014.
2. Nigel Cameron, M. Ellen Mitchell, John Wiley & Sons, 2007.
3. Sylvia Leydecker, Marius Kölbel, Sascha Peters, Springer Science & Business Media, 2008.
4. Sharon Maheshwar, Academy Publish, 1969.
5. W. Kulisch, R. Freudenstein, A. Ruiz, A. Valsesia, L. Sirghi, J. Ponti, P. Colpo, F. Rossi, Springer Science, 2009.
6. M.H. Fulekar, I. K. International Pvt Ltd, 2010.
7. Christian Ngo, Marcel van de Voorde, Springer Science & Business Media, 2014.
8. Jitendra N. Tiwari, Rajanish Tiwari, Kwang Kim, Progress in Materials Science 57 (2012) 724-803.
9. Manijeh Razeghi, Springer US, (2002) 349-386.
10. Geoffrey Strouse, Jeffrey Gerbec, Donny Magana, The regents of the university of California, 2006.
11. Clemens Burda, Xiaobo Chen, Radha Narayanan, Mostafa A. El-Sayed, Chemical Reviews 105 (2005) 1025-1102.
12. Shirasaki, Supran, Bawendi, Bulovic, Nature Photonics 7 (2013) 13-23.
13. R. Rossetti, S. Nakahara, L. E. Brus, Journal of Chemical Physics 79 (1983) 1086-1088.
14. A. I. Ekimov, A. A. Onushchenko, Soviet physics: Semiconductors 16 (1980) 775-778.
15. A. I. Ekimov, A. A. Onushchenko, Semiconductors 16 (1982) 1215-1219.
16. Ronny Costi, Aaron E. Saunders, Uri Banin, Angewandte Chemie International Edition 49 (2010) 4878–4897.
17. E. Matijevi, D. M. Wilhelmy, Journal of Colloid and Interface Science 86 (1982) 476-484.
18. L. Spanhel, M. Haase, H. Weller, A. Henglein, Journal of the American Chemical Society 109 (1987) 5649-5655.
19. Colvin, Schlamp, Alivisatos, Nature 370 (1994) 354-357.
20. Schlamp, Peng, Alivisatos, Journal of Applied Physics 82 (1997) 5837-5842.
21. M. C. Schlamp, University of California, Berkeley, 1998.
22. Greenham, Peng, Alivisatos, Physical Review B 54 (1996) 17628-17637.
23. Greenham, Peng, Alivisatos, Synthetic Metals 84 (1997) 545-546.
24. Klein, Roth, Lim, Alivisatos, McEuen, Nature 89 (1997) 699-701.
25. Alivisatos, MRS Bulletin 23 (1998) 18-23.
26. Bruchez, Moronne, Gin, Weiss, Alivisatos, Science 281 (1998) 2013-2016.
27. Chan, Nie, Science 281 (1998) 2016-2018.
28. Surender Kumar, Tukaram J. Shinde, Pramod N. Vasambekar, Advanced Materials Letters 4 (2013) 373-377.
29. Nehar Celikkin, Lucie Jakubcova, Martin Zenke, Mareike Hoss, John Erik Wong, Thomas Hieronymus, Journal of Magnetism and Magnetic Materials 380 (2015) 39-45.
30. R J Tackett, J G Parsons, B I Machado, S M Gaytan, L E Murr, Botez, Nanotechnology 21 (2010) 365703.
31. Arati Kolhatkar, Andrew Jamison, Dmitri Litvinov, Richard Willson, Randall Lee, International Journal of Molecular Sciences 14 (2013) 15977-16009.
32. M.Todd, Louise Prakash, Satya Prakash, Cell press 107 (2001) 917-927.

33. Xiaosheng Fang, Yoshio Bando, Ujjal K. Gautam, Tianyou Zhai, Haibo Zeng, Xijin Xu, Meiyong Liao, Dmitri Golberg, Critical Reviews in Solid State and Materials Sciences, 34 (2009) 190-223.
34. Dandan Lin, Hui Wu, Rui Zhang, Wei Zhang, Wei Pan, Journal of the American Ceramic Society 93 (2010) 3384-3389.
35. Jin YJ, Duan XB, Lin YM, Wang SK, Journal of Guang Pu Xue Yu Guang Pu Fen Xi 32 (2012)1278-1281.
36. P. K. Nair, V. M. Garcia, A. M. Fernandez, H. S. Ruiz, and M. T. S. Nair, Journal of Physics D 24 (1991) 441-449.
37. A. A. Sagade and R. Sharma, Sensors and Actuators B 133 (2008) 135-143.
38. Azam sobhani, Masoud salavati-Niasari, Superlattices and Microstructures 59 (2013) 1-12.
39. Sohail Saeed, Naghmana Rashid, Saeed & Rashid, Cogent Chemistry 1(2015) 1030195.
40. Smotkin, Cervera-March, Bard, Campion, Fox, Mallouk, Webber, White, Journal of Physical Chemistry 91 (1987) 6-8.
41. S. Mochizuki, B. Piriou, Journal of Physics: Condensed Matter 2 (1990) 5225.
42. T. Dietl, H. Ohno, F. Matsukura, J. Cibert, D. Ferrand, Science 287 (2000) 1019.
43. R. Tappero, P. D'Arco, A. Lichanot, Chemical Physics Letters 273 (1997) 83-90.
44. D. Hobbs, J. Hafner, Journal of Physics: Condensed Matter 11 (1999) 8197-8222.
45. A.N. Kravtsova, I.E. Stekhin, A.V. Soldatov, X. Liu, Physical Review B 69 (2004) 134109.
46. R.L. Gunshor, A.V. Nurmikko, Semiconductors and Semimetals 44 (1997) 163.
47. N. Zhang, R. Yi, Z. Wang, R. Shi, H. Wang, G. Qiu, X. Liu, Materials Chemistry and Physics 111 (2008) 13-16.
48. M. Girish, T. Dhandayuthapani, R. Sivakumar, C. Sanjeeviraja, Asian Journal of Applied Sciences 7 (2014) 729-736.
49. Ian T. Sines, Rajiv Misra, Peter Schiffer, Raymond E. Schaak, Angewandte Chemie International Edition 49 (2010) 4638-4640.
50. Tom A. Bither, Wilmington, Patented March 12 (1968) US3372997.
51. Thomas Weber, Roel Prins, Rutger A. van Santen, Springer Netherlands, 1998.
52. Kjemisk Institutt A, Universitetet Oslo, Blindern, Acta Chemic Scandinavica 19 (1965) 1405-1410.
53. Carolyn I. Pearce, Richard A D Pattrick, David J. Vaughan, Reviews in Mineralogy and Geochemistry 61 (2006) 127-180.
54. Subhajit Biswas, Soumitra Kar, Subhadra Chaudhuri, Materials Science and Engineering B 142 (2007) 69-77.
55. Cecil J. Schneer, Richard W. Whiting, American Mineralogist, 48 (1963) 737-758.
56. Tian Xi Wang, Wei Wei Chen, Chemical engineering journal 144 (2008) 146-148.
57. D Hobbs, J Hafner, Journal of Physics: Condensed Matter 11 (1999) 8197-8222.
58. Ning Zhang, Ran Yi, ZhongWang, Rongrong Shi, HaidongWang, Guanzhou Qiu, Xiaohe Liu, Materials Chemistry and Physics 111 (2008) 13-16.

59. Veeramanikandasamy T, Rajendran K, Sambath K, Asian Journal of Chemistry 25 (2013) S55-S58.
60. Elena A Trusova, Kirill V Vokhmintcev, Igor V Zagainov, Nanoscale Research Letters 7 (2012) 58-62.
61. Steven C. Erwin, Lijun Zu, Michael I. Haftel, Alexander L. Efros, Thomas A. Kennedy, David J. Norris, Nature 436 (2005) 91-94.
62. Jesse H. Engel, Yogesh Surendranath, A. Paul Alivisatos, Journal of the American Chemical Society 134 (2012) 13200-13203.

Chapter II
REVIEW OF LITERATURE

2.1 Introduction

In recent years, the synthesis of new semiconductor nanomaterials with special properties remains one of the major challenges in the field of material science. Complexity of the field of semiconductor materials with special properties is due to their interdisciplinary characteristics which combining different aspects of chemistry, physics and engineering. Currently, research in this field is a continuous challenge and knows a strong revival due to the multiple applications of these materials in opto-electronics and magneto-optical devices. The synthesis of colloidal semiconductor nanocrystals (NCs) has been rapidly developing research area with a particular emphasis on II-VI semiconductor systems [1]. The special focus is on the literature published on the synthesis, characterization and applications of undoped and doped manganese sulfide powder as well as thin film have been reported. This will help the reader to get an idea of the importance of work have been performed in the present study. This work has led to significant improvements in particle size tunability, reproducibility, production of particles with superparamagnetism, and adjustment of optical and electronic properties.

However, further improvements in material quality and monodispersity are still possible and desirable in all of these areas (viz. for improvement in the performance of NC-based devices). More importantly, continued research on the synthesis of MnS NCs will yield a more complete understanding of this important model system, and this improved understanding will form the basis for synthetic optimization of a broader range of semiconductor NCs.

2.2 Survey of Previous Work

The manganese chalcogenides are magnetic materials that have unique physical, morphological and chemical properties [2-4]. Manganese chalcogenides have attracted considerable interest due to their intriguing properties and structural diversity. The small dimensions of these chalcogenide nanocrystals result in different physical properties from those observed in the corresponding macrocrystalline bulk materials. As particle sizes become smaller the ratio of surface atoms to the interior increases, leading to the surface properties playing a vital role in the properties of the materials. Chalcogenide nanocrystals also exhibit a change in their electronic properties relative to those of the bulk materials. This allows chemist and materials scientists to change the electronic properties of the materials simply by controlling their sizes; progress in the synthesis of chalcogenide nanoparticles will promote the research on their applications. Meanwhile, advances in the applications will bring new challenges in the synthesis of these materials to both synthetic chemists and materials scientists. The manganese sulfide is a wide band gap ($Eg \approx 3.8eV$) diluted magnetic p-type semiconductors [5]. The wide-bandgap chalcogenide semiconductors are of current interest for optoelectronic application, light-emitting diodes and optical devices. There are various techniques to synthesize MnS nanocrystals such as solvothermal synthesis [6], successive ionic layer adsorption and reaction (SILAR) [7],

Hydrothermal method [8, 9] and Chemical bath deposition. This part is an integral part of continued investigation of manganese sulfide. It deals with various aspects of synthesis and characterization.

2.2.1 MnS Thin Films

M. Dhanam et al. [10] have prepared γ-MnS thin films comprising nanoparticles by a chemical bath deposition technique. Solution of manganese acetate, triethylamine and ammonia were mixed slowly in deionized water at room temperature. A glass substrate was mounted vertically in a beaker containing this solution. Hydrazine hydrate and thioacetamide solution was added at 80°C. XRD analysis is employed to confirm the structure and nature of the nanocrystals. The calculated crystallite size ranges from 40-54 nm. SEM pictures showed the hollow spherical morphology of the prepared γ-MnS thin films. **DongBo Fan et al.** [11] prepared the pure γ-MnS films with pre-treatment substrates by HF etching and control of the reaction temperature play an important role in the formation of crystalline films. The optical band gaps of as-deposited amorphous and γ-MnS films are reported as 3.23 and 3.16 eV, respectively.

Anuar Kassim et al. [12] have obtained the MnS thin films onto indium tin oxide glass substrate. Manganese sulphate and thiourea were used as precursor which supplied Mn^{2+} and S^{2-} ions. X-ray diffraction and atomic force microscopy were used to investigate the structural and morphological properties of films, respectively. They also reported that the band gap of the sample is dependent on the bath temperature. **Gajanan Pandey et al.** [13] have synthesized γ-MnS nanocrystalline materials by the reaction of $Na_2[Mn(HL)_2(H_2O)_2]$; 1:2 (M:L) chelate complex with alkaline solution of thiocarbamide in aqueous solution phase. Effect of metal chelate complex, reaction time and surfactant sodium dodecyl sulfate on phase, morphology and size of the products have been investigated.

C. Gumus et al. [14] prepared highly transparent crystalline manganese sulfide (γ-MnS) thin films by CBD method at room temperature. The refractive index, extinction coefficient and film thickness are evaluated using envelope method. The optical band gap of the film was estimated to be 3.88 eV. The effect of trisodium citrate on crystalline films has been determined by them. **DongBo Fan et al.** [15] studied the photoluminescence of the γ-MnS film from 295K down to 30K and exhibited two emission bands centered at 1.8 and 1.66 eV. The temperature dependence of photoluminescence was also observed and interpreted using the spin-wave assist photoluminescence model.

YongCai Zhang et al. [16] successfully synthesized phase-pure metastable γ-MnS crystallites via hydrothermal reactions between manganese acetate $(Mn(CH_3COO)_2)$ and thioacetamide in distilled water at 60-130 °C for 20 h. It was found that the phase(s) of as-formed MnS crystallites depended on the reaction temperature and time. **D. Sreekantha Reddy et al.** [17] have formed MnS nanocrystalline films on glass substrates by thermal evaporation technique at room temperature and the films were annealed at 573 K. The films were characterized by using EDAX, XRD, SEM and AFM. The grain size is in the range between 30 - 32 nm and exhibited wurtzite structure. The lattice parameter and band gaps (3.842 – 3.916 eV) increase with increasing annealing temperature.

H.M. Pathan et al. [7] have deposited MnS thin films by a simple and inexpensive successive ionic layer adsorption and reaction (SILAR) method. The as-deposited film on glass substrate was amorphous. As thickness increases optical band gap was found to be increase. The water angle contact was found to be 348, suggesting hydrophilic nature of manganese sulfide thin films.

Sandra A. Mayen-Hernandeza et al. [18] have obtained polycrystalline MnS thin films by the RF-sputtering technique above room temperature for the first time.

The addition of sulfur to the targets for growing films at room temperature leads to amorphicity (5 at%) or to a change in the crystal orientation of the films (10 at%). A bandgap energy value of 3.3 eV was found for γ-MnS films. **Y.M. Yu et al.** [19] have investigated the epitaxial growth characteristics of α-MnS on GaAs (1 0 0) substrates by X-ray diffraction and double crystal rocking curve measurements. The films on GaAs at low substrate temperature exhibit multiphase crystal structures of zincblende and rock-salt, and it is changed to rock-salt with increasing substrate temperature.

Yong Shi et al. [20] have deposited pure metastable β-MnS thin films by CBD method and subsequently annealed in a Na_2S solution. The as-deposited β-MnS can be transformed to γ-phase MnS after the hydrothermal treatment in an autoclave at 200 °C for 1 h. The estimated Eg values are in the range of 3.15–3.18 eV. **C. Gumus et al.** [21] have deposited crystalline γ-MnS thin film on glass substrate at room temperature by using CBD method. The magnetic structure of prepared film was characterized by means of several parameters including magnetic moment and magnetic susceptibility. The magnetic behavior is an antiferromagnetic Neel type.

M.N. Nnabuchi [22] has deposited six thin films of manganese sulfide using solution growth technique at different bath parameters, which include temperature, molarity of solution, volume of solution and water, time of deposition and pH. The bath compositions are Manganese chloride ($MnCl2.4H2O$), sodium thiosulphate ($Na2S2O3$), distilled water, and ethylene diamine tetra acetic acid (EDTA) which served as the complexing agent. The optical and solid state characteristics revealed that films of manganese sulfide (MnS) have low absorbance ~0.033-0.40, high transmittance ~4-99%, and low reflectance ~0-20% throughout the ultraviolet, visible, and infrared regions. The thickness t is ~1.96-2.69μm, refractive index n is ~1.39-1.74 and energy band gap E_g is ~2.60-3.90eV.

The above results show that MnS could be coated on solar collectors to enhance solar energy collection. It could also be used as anti-reflection coatings.

J. Yuvaloshinia et al. [23] have prepared ZnS/MnS super lattice thin films deposited on glass substrates by Chemical Bath Deposition Technique. They have analyzed the effect of annealing on optical and structural properties of ZnS/MnS. The MnS/ZnS superlattices thin films could be used in solar energy application.

Feng Tao et al. [24] have synthesized the complex three dimensional (3D) α-MnS flowerlike hierarchical architectures with 1-2 mm in diameter ultrathin nanosheets, with 600 nm in length, 150-300 nm in width, and about 10-20 nm in thickness by a simple hydrothermal method. Selected area electron diffraction (SEAD) pattern and high resolution transmission electron microscope (HRTEM) image indicate that the nanosheet is a well-developed single crystal. A clear perspective is shown that more complex nanostructured material could be chemically synthesized.

2.2.2 MnS Powders

Changhua et al. [9] successfully developed α-MnS nanorods under hydrothermal conditions using manganese and sulfur powder as starting materials in the temperature range of 240-260 °C. It was found that the temperature was a key factor influencing the phase purity of the α-MnS nanorods. **Pingtang Zhao et al.** [8] have synthesized MnS hollow spheres consisting of cones by a dodecanethiol-assisted hydrothermal process at 180 °C for 24 h, employing manganese acetate tetrahydrate and L-cysteine as the precursors. The diameter of the γ-MnS hollow spheres is 3-6 mm. The UV–Vis spectrum of the γ-MnS hollow spheres has peak at about 279 nm, and the fluorescence spectrum of the γ-MnS hollow spheres has two emission peaks at 424 and 560 nm.

Jianming Huang et al. [25] group has synthesized oriented MnS nanorods on anodic aluminum oxide template under a hydrothermal condition and demonstrated the effect of precursor content on the morphology of samples. **Meiying Liu et al.** [26] have successfully produced the γ-MnS multipods consisting of hexagonal prism-like rods by a simple bio molecule-assisted hydrothermal approach using l-cystine as both sulfur source and complexing agent. The room-temperature photoluminescence spectrum exhibits a peak centered at 419 nm corresponding to the band edge emission when the sample was excited with a wavelength of 358 nm.

F. M. Michel et al. [27] report a hydrothermal synthesis method for the formation of MnS in which a $MnCl_2$ solution is injected into a preheated sulfide solution. By varying the temperature of injection and subsequent aging time controls the specific crystal phase of the product. Both α-MnS and γ-MnS have been prepared by varying the process temperature. **Zhijun Wang et al.** [28] MnS flower-like hierarchical architectures were successfully prepared on the surface of the porous alumina membrane (PAM) under hydrothermal condition. PAM is indispensable for forming MnS nanowires and gas bubbles formed within the nanopores of PAM by the thiourea decomposition play an important role for the self-assembled MnS flower-like architectures. The room-temperature PL spectrum shows a strong emission peak at 420 nm corresponding to the MnS band edge emission.

Subhajit Biswas et al. [29] have synthesized MnS micro and nanocrystals with different morphologies such as pyramid, octahedron, sheet, sphere and hexagonal ring by a solvothermal route. It was observed that the experimental parameters such as the variation in the precursor salt, molar ratio of the anionic and cationic precursors as well as the solvent play critical role in determining the morphology and crystal structure of the products. Interpenetrating crystal growth and clustering formation were also observed under certain circumstances.

Jin Mu et al. [30] have prepared MnS nanocrystals with different phases and shapes through solvothermal synthesis. The results showed that the solvent and reaction time played a critical role in controlling the phase and shape of MnS nanocrystals. The mixture of α, β and γ-MnS was obtained in water at 5 h and the palm-like structure was observed.

Tian XiWang et al. [31] were synthesized pure rock-salt structure manganese sulfide (α-MnS) powders via solvothermal decomposition of an easily obtained single-source molecular precursor (manganese N, N'-diethyl dithiocarbamate: Mn-DDTC) in 80 vol.% hydrazine hydrate aqueous solution at 90-120 °C for 24 h. Mn-DDTC, the single-source molecular precursor with pre-established Mn–S chemical bonds, played a key role in preparing pure α-MnS under the current solvothermal conditions. **T. Kurz et al.** [32] have studied the paramagnetic-to-antiferromagnetic phase transition in spherical β-MnS nanoparticles of well defined diameters in the range of 3-11 nm. The MnS nanoparticles were obtained by intra-pore synthesis inside mesoporous silica matrices. Electron spin resonance and magnetization measurements reveal that no antiferromagnetic order is established in MnS spheres of 3 nm down to 2 K and that the antiferromagnetic order is gradually recovered on increasing the particle diameter to 11 nm.

Yange Zhang et al. [33] have synthesized the stable α-MnS by solvothermal method in different solutions at relatively low temperature (120 °C for 12 h). The effect of the solvents, surfactant, reaction temperature and reaction time on the morphology and phase of products has been investigated. This microcrystals and nanocrystals might have potential applications in micro devices, magnetic cells and optical materials among others. This synthesis method could be extended to prepare diluted magnetic semiconductors (DMSs). **Xinhua Zhang et al.** [6] successfully synthesized the stable α-MnS spheres by two-step solvothermal method and traditional solvothermal technique without any surfactant assistance.

This two-step synthesis helps to know the growth mechanism in solvothermal synthesis is better.

Sunil H. Chaki et al. [34] have synthesized the MnS nanoparticles at ambient temperature by wet chemical technique. Manganese acetate ($C_4H_6MnO_4 \cdot 4H_2O$) was used as source for Mn^{+2} ions and thioacetamide (C_2H_5NS) was used as source for S^{-2} ions. The crystallite size calculated from XRD using Scherrer's formula and Hall-Williamson relation came out to be of 6.81 nm and 5.27 nm respectively. The optical absorption spectra showed absorption edge at 325 nm corresponding to energy of 3.82 eV, which acknowledged the occurrence of blue shift. The photoluminescence spectra recorded for five different excitation wavelengths viz 250, 275, 280, 300 and 325 nm showed three emission peaks at 463 nm, 550 nm and 821 nm.

Long Peng et al. [35] have synthesized MnS nanocrystals with well-defined shapes and crystal structures such as hexapod, octahedral, hexagonal shaped α- MnS NCs, and pencil-shaped γ-MnS NCs by facile single-source precursor method. The effects of the composition of precursor, reaction temperature, and the heating rate on the morphologies, and crystal structures of MnS NCs were systematically studied for the first time. The IR reflection measurements of α-MnS have been performed at room temperature under various pressures by **Y. Ishidaa et al.** [36]. **Ning Zhang et al.** [5] have prepared stable rock-salt type α-MnS sub-microcrystals by hydrothermal synthetic route under mild conditions. The results revealed that the electrochemical performance of the α-MnS sub-microcrystals may be associated with the degree of crystallinity and particle size of samples. The initial lithiation capacity of the α-MnS sub-microcrystals obtained at 120 °C is 1327mAhg^{-1} at 0.7V versus Li/Li$^+$, which exhibited α-MnS sub-microcrystals is extremely promising anode material for lithium-ion batteries and has great potential applications in the future.

Yang Ren et al. [37] have prepared well-crystallized γ-MnS hierarchical nanostructures with different morphologies and particle sizes by a solvothermal method in the ethylene glycol–H_2O system. Accordingly, nanorod constructed nanoflowers and hollow nanospheres can be selectively prepared in the presence and absence of deionized water, respectively. As for the formation of γ-MnS nanoflowers, a synergistic growth mechanism combined oriented attachment followed by Ostwald ripening has been proposed by taking into account the kinetic inducing effect of deionized water. Moreover, increasing the dosage of the deionized water would increase the crystallinity and particle size of γ-MnS nanocrystals, and up to a certain amount, phase transformation to a phase occurs. The obtained γ-MnS nanorod-constructed nanoflowers exhibit improved PL property compared to hollow nanospheres, evidenced by the enhanced NBE emission centered at 364 nm.

H.H. Heikens et al. [38] have reported the thermoelectric power, Hall Effect and resistivity on iodine-grown crystals of P-type α-MnS. A study of the temperature dependences reveals that the conductivity occurs by holes in a 3D-band (Mn^{3+}). **N. Moloto et al.** [39] have synthesized manganese chalcogenide nanoparticles using mild synthetic methods. The MnS nanoparticles were synthesized using a single-source precursor method with tetramethylthiuram disulfide as the ligand, whilst the MnSe nanoparticles was synthesized using a modified hydrothermal method. MnS nanoparticles with wire-like morphology with high aspect ratio and MnSe nanorods that were nearly mono-dispersed were reported. The optical properties of both MnS and MnSe showed the quantum confinement effects. Furthermore, both MnS and MnSe showed paramagnetic characteristics with the ESR spectra showing broad single resonance peaks with g-values of 2.0064 and 2.0068 respectively. The temperature dependence of resistivity, magnetization and electron-spin resonance of the α-MnS single crystal were measured in temperature range of $5K < T < 550K$.

Magnetization hysteresis in applied magnetic field up to 0.7 T at T = 5K, 77K, 300K, irreversible temperature behavior of magnetization and resistivity were found.

The contribution of holes in manganese ions to the conductivity, optical absorption spectra, charge susceptibility maxima and charge instability in α-MnS were studied by **S.S. Aplesnin et al.** [40]. **Qiwei Tian et al**. [41] have developed an effective method to control growth of well-defined star-shaped α-MnS nanocrystals via a cooperative thermal decomposition route using the mixed surfactant solution. These star-shaped α-MnS nanocrystals show novel magnetic properties, i.e., a high blocking temperature (275 K) and a large coercive field (1573 Oe) which is due to the main interaction between the spins of the core (antiferromagnetic) and the spins of the shell (insulating and disordering).

Dekun Ma et al. [42] have successfully synthesized α-MnS single crystalline nanobelts from the reaction of manganese acetate with ammonium thiocyanate in dodecylamine by one-pot solvothermal route. The slow release of sulfur ions and the formation of manganese-dodecylamine complex is the key issue for the nucleation and growth of high-quality α-MnS nanobelts. The study of magnetic properties of α-MnS nanobelts shows that nanobelts exhibit a higher Neel temperature (35 K) and larger coercivity (4013 Oe). The study on the electrical transport properties of individual α-MnS nanobelt has a conductivity of 3.2-105 S/m. The unique magnetic properties together with semiconducting characteristics of these α-MnS nanobelts make them promising potential applications in nanodevices. **S.S.Aplesnin et al.** [43] have studied the resistivity, the optical absorption spectra of single crystal α-MnS in the temperature range 80–300 K along two directions [100] and [111].

Strong anisotropy of resistivity, shift of absorption spectra band edge below T, 160 K are explained in terms of model delocalized holes in 3d-band manganese ions interacting with localized spins by using sd-model.

Felicia Iacomi et al. [44] have synthesized well-defined MnS clusters within the pore structure of zeolites clinoptilolite and laumontite. XRD and magnetic resonance measurements proved that the clusters are formed inside the zeolite channels. The partial vacancies and imperfections introduce inhomogeneity in the cluster size distribution, which is the reason for the broad excitonic absorption peak and the broad luminescence spectra. The NMR spectra indicate that structural changes occurred in the tetrahedral framework during the sulfidation process. EPR measurements performed at room temperature evidenced three different Mn species which are affected by the sulfidation process [45].

2.3 Motivation of the Present Work

The nanomaterials may be metallic, semiconductors, polymer, ceramic or glass. Amongst the various types of nanomaterials, the semiconductor nanoparticles have been widely investigated. A gradual transition from solid state to molecular structure occurs as the particle size decreases in the nanoscale structure. In this very size regime, the physical and chemical properties of the particles strongly depend on their size. Nanocrystalline semiconductors offer a wide variety of possible applications in nanoelectronics, nanophotonics, optical and optoelectronic devices, LED's, resonators in wave guide, solar cells, biosensors, gas sensors, lithium-ion batteries, gamma-ray detectors, TFTs, PEC cells, etc. The synthesis of binary metal chalcogenides semiconductors in a nanocrystalline form has been a rapidly growing area of research due to their novel electronic, magnetic, optical, and structural properties, quantum size effect and other important physical and chemical properties.

Particularly the MnX, ZnX, and PbX (X: Sulfides, Selenides and Tellurides) have received considerably more attention. During the past few decades, manganese chalcogenides (MnS, MnSe, MnTe) are a category of magnetic materials that have attracted interests concerning their structural, chemical and physical properties.

Among these manganese compounds, MnS belongs to the family of diluted magnetic semiconductor (DMS) and has been extensively studied because of outstanding magneto-optical properties. In DMS, the band electrons and holes strongly interact with the localized magnetic moments and cause a variety of interesting phenomena. The MnS is a wide band gap (E_g = 3.1 eV) p-type magnetic semiconductor material having potential applications in solar cell as window/buffer material, solar selective coatings, short wavelength optoelectronic materials, sensors, photoconductors, optical mass memories and storage. It is a typical direct band gap semiconductor, which has been a model material in the studies of quantum size effects.

Wide bandgap semiconductor nanocrystals have attracted the scientific community in the past decades because of their size dependent properties and diverse applications. The bandgap energy can be tuned by slight variation in size and composition, which enables them to be used in variety of applications with an increase in efficiency. A simultaneous control of structure and morphology of these nanocrystals provides opportunities to tune and explore their optical properties. Manganese sulfide is a VIIB-VIA group semiconductor which MnS has three crystalline forms: thermodynamically stable α-MnS phase with an octahedrally coordinated rock salt structure, pink metastable tetrahedrally coordinated β and γ-MnS with zincblende and wurtzite structures respectively. It has been found that β and γ-MnS can transform irreversibly to the stable α-MnS in the range of 373 to 673 K.

In previous studies, some considerable efforts in the field have been placed on the control of the crystalline phases and morphologies of manganese sulfide thin films and nanocrystalline powders. To the best of our knowledge, very few electrical and magnetic properties of manganese sulfide have been investigated so far, although some other metal sulfides have been widely studied for many years.

This is the underlying motivation for the electrical and magnetic characterization of the prepared samples. Besides its own importance as a semiconductor and magnetic material, MnS is also linked to the study of semimagnetic alloys that contain Mn and S. Indeed, current investigations in the field of diluted magnetic semiconductors (DMS) make the full knowledge of the properties of the binary end compounds of such ternary mixed crystals desirable (Cd-MnS, Zn-MnS). The novel magneto-optical phenomena in these materials have provoked the exploration about the synthesis of magnetic ions doped nanocrystals (NCs). Motivated by these finding, we synthesized undoped MnS nanocrystals with two controlling parameters (processing temperature and molar ratio) and Cu-doped MnS nanocrystals, and investigated their structural, optical, electrical and magnetic properties.

The literature reveals that the various methods so far adopted for the preparation of MnS nanocrystals, such as thermal vacuum evaporation, radio frequency sputtering, microwave irradiation, chemical bath deposition (CBD), hydrothermal, solvothermal, microwave irradiation, chemical vapor deposition techniques. However, there are not much previous reports available on the preparation of MnS nanocrystalline powder by wet chemical route without using sophisticated instruments.

2.4 Objective of the Present Work

As an important I–VII semiconductor material, Manganese Sulfide (MnS) is a wideband gap p-type diluted magnetic semiconductor, hence it is considered to be a promising host material. The objective of the study consists in determining the structural, morphological, optical, electrical and magnetic properties of manganese sulfide nanocrystals with different process parameters and cu-doped manganese sulfide nanocrystalline powders prepared by wet-chemical synthesis at low temperature.

To control the particle size, morphology, electrical and magnetic properties by adjusting the process parameters. The synthesized samples were characterized by using X-ray diffraction (XRD), Transmission Electron Microscope (TEM), Scanning Electron Microscope (SEM), Fourier Transforms Infrared Spectroscopy (FTIR), UV-Visible absorption and Photoluminescence (PL) Spectroscopy, LCR Hi-Tester meter and Vibrating Sample Magnetometer (VSM) techniques.

The investigations have pursued to highlight the influence of following process parameters:

1. Influence of refluxing temperature on the structural, optical, electrical and magnetic properties of chemically synthesized MnS nanocrystals

    A first part of this research results were published in Asian Journal of Chemistry, ISSN 0970-7077, Volume-25, 2013, ppS55-S58.

2. Influence of Mn/S molar ratio on the structural, optical, electrical and magnetic properties of chemically synthesized MnS nanocrystals

    A second part of this research results were published in Journal of Materials Science: Materials in Electronics, ISSN 0957-4522, Volume-25, Issue-8, 2014, pp3383-3388.

3. Influence of Cu-doping on structural, optical, electrical and magnetic properties of chemically synthesized MnS nanocrystals

A third part this research results is submitted in Materials Chemistry and Physics, ISSN: 0254-0584 (Under review).

2.5 Summary of the Contributions

There are three main contribution of this research work,

1. The β-MnS and γ-MnS nanocrystalline powders using simple wet chemical technique at low temperature range have been synthesized.
2. The influences of process parameters on electrical properties like dielectric, impedance, electric modulus and AC conductivity analysis have been recorded.

The structural, optical, electrical and magnetic properties of cu-doped MnS nanocrystals have been studied.

References

1. Rolf Koole, Esther Groeneveld, Daniel Vanmaekelbergh, Andries Meijerink, Celso de Mello Donega, Springer-Verlag Berlin Heidelberg 2014.
2. T. Mahalingama, S. Thanikaikarasana, V. Dhanasekarana, A. Kathalingamb, S. Velumanic, Jin-Koo Rheeb, Materials Science, Engineering B 174 (2010) 257-262.
3. Raj Kishore Sharmaa, Gurmeet Singhb, Yong Gun Shula, Hansung Kima, Physica B 390 (2007) 314-319.
4. S.H. Wei, A. Zunger, Physical Review B 48 (1993) 6111.
5. Ning Zhang, Ran Yi, ZhongWang, Rongrong Shi, HaidongWang, Guanzhou Qiu, Xiaohe Liu, Materials Chemistry and Physics 111 (2008) 13-16.
6. Xinhua Zhang, Yiqing Chen, Chong Jia, Qingtao Zhou, Yong Su, Bo Peng, Song Yin, Minjun Xin, Materials Letters 62 (2008) 125-127.
7. H. M. Pathan, S. S. Kale, C.D. Lokhande, S.H. Han, O.S. Joo, Materials Research Bulletin 42 (2007) 1565-1569.
8. Pingtang Zhao, Qiumei Zeng, Xianliang He, Hao Tang, Kaixun Huang, Journal of Crystal Growth 310 (2008) 4268-4272.
9. Changhua An, Kaibin Tang, Xianming Liu, Fanqing Li, Guien Zhou, Yitai Qian, Journal of Crystal Growth 252 (2003) 575-580.
10. M. Dhanam, B. Kavitha, M. Shanmugapriya, Chalcogenide Letters 6 (2009) 541- 547.
11. DongBo Fan, Hao Wang, YongCai Zhang, Jie Cheng, Bo Wang, Hui Yan, Materials Chemistry and Physics 80 (2003) 44-47.
12. Anuar Kassim, Ho Soon Min, International Journal of Chemical Research 1 (2010) 1-5.
13. Gajanan Pandey, Harendra K. Sharma, S.K. Srivastava, R.K. Kotnala, Materials Research Bulletin 46 (2011) 1804-1810.
14. C. Gumus, C. Ulutas, Y. Ufuktepe, Optical Materials 29 (2007) 1183-1187.
15. DongBo Fan, XiaoDong Yang, Hao Wang, YongCai Zhang, Hui Yan, Physica B 337 (2003) 165-169.
16. YongCai Zhang, Hao Wang, Bo Wang, HaiYan Xu, Hui Yan, Masahiro Yoshimura, Optical Materials 23 (2003) 433-437.
17. D. Sreekantha Reddy, D. Raja Reddy, B. K. Reddy, A. Mallikarjuna Reddy, K. R. Gunasekhara, P. Sreedhara Reddy, Journal of optoelectronics and advanced materials 9 (2007) 2019-2022.
18. Sandra A. Mayen-Hernandeza, Sergio Jimenez-Sandovala, Rebeca Castanedo-Pereza, Gerardo Torres-Delgadoa, Benjamin S. Chaob, Omar Jimenez-Sandovala, Journal of Crystal Growth 256 (2003) 12-19.
19. Y.M. Yu, D.J. Kim, Y.D. Choi, C.S. Kim, Applied Surface Science 253 (2007) 3521-3524.
20. Yong Shi, Fanghong Xue, Chunyan Li, Qidong Zhao, Zhenping Qu, Materials Research Bulletin 46 (2011) 483-486.
21. C. Gumus, A. Bayri, C. Uluta, M. Karakaplan, Y. Ufuktepe, Journal of optoelectronics and advanced materials 8 (2006) 261-265.
22. M.N. Nnabuchi, The Pacific Journal of Science and Technology, Spring 7 (2006) 69-76.
23. J. Yuvaloshinia, Ra. Shanmugavadivu, G. Ravi, Optik 125 (2014) 1775-1779.
24. Feng Tao, Zhijun Wang, Lianzeng Yao, Weili Cai, Xiaoguang Li, Materials Letters 61 (2007) 4973-4975.

25. Jianming Huang, Weifeng Liu, Yong L, Lianzeng Yao, Bulletin of the Korean Chemical Society 31 (2010) 2447-2448.
26. Meiying Liu, Nannan Shan, Linlin Chen, Xiaoqian Li, Bona Li, Wansheng You, Applied Surface Science 258 (2012) 7922-7927.
27. F. M. Michel, M. A. A. Schoonen, X. V. Zhang, S. T. Martin, J. B. Parise, Chemistry of Materials 18 (2006) 1726-1736.
28. Zhijun Wang, Feng Tao, Feng Pan, Yufeng Sun, Weili Cai, Lianzeng Yao, Applied Surface Science 258 (2011) 44-49.
29. Subhajit Biswas, Soumitra Kar, Subhadra Chaudhuri, Materials Science and Engineering B 142 (2007) 69-77.
30. Jin Mu, Zhenfang Gu, Lei Wang, Zhiqing Zhang, Hua Sun, Shi-Zhao Kang, Journal of Nanoparticle Research 10 (2008) 197-201.
31. Tian XiWang, WeiWei Chen, Chemical Engineering Journal 144 (2008) 146-148.
32. T. Kurz, L. Chen, F. J. Brieler, P. J. Klar,H.A. Krug von Nidda, M. Fröba,W. Heimbrodt, A. Loidl, Physical Review B 78 (2008) 132408.
33. Yange Zhang, Zude Zhang, Shutao Wang, Xuchu Ma, Yitai Qian, Materials Chemistry and Physics 97 (2006) 365-370.
34. Sunil H. Chaki, M.P. Deshpande, J.P. Tailor, K.S. Mahato, M.D. Chaudhary, Advanced Materials Research 584 (2012) 243-247.
35. Long Peng, Shuling Shen, Yejun Zhang, Huarui Xu, Qiangbin Wang, Journal of Colloid and Interface Science 377 (2012) 13-17.
36. Y. Ishidaa, Y. Mitaa, M. Kobayashia, S. Endob, S. Mochizukic, Journal of Magnetism and Magnetic Materials 272 (2004) 428-429.
37. Yang Ren, Lian Gao, Jing Sun, Yangqiao Liu, Xiaofeng Xie, Ceramics International 38 (2012) 875-881.
38. H.H. Heikens, C.F. Van Bruggen, C. Haas, Journal of Physics and Chemistry of Solids 39 (1978) 833-840.
39. N. Moloto, M.J. Moloto, M. Kalenga, S. Govindraju, M. Airo, Optical Materials 36 (2014) 31-35.
40. S.S. Aplesnin, L. I. Ryabinkina, G. M. Abramova, O.B. Romanova, A.M. Vorotynov, D.A. Velikanov, N. I. Kiselev, A. D. Balaev, arXiv:cond-mat/0408232, 2004.
41. Qiwei Tian, Minghua Tang, Feiran Jiang, Yiwei Liu, Jianghong Wu, Rujia Zou, Yangang Sun, Zhigang Chen, Runwei Li, Junqing Hu, Chemical Communications 47 (2011) 8100-8102.
42. Dekun Ma, Shaoming Huang, Lijie Zhang, Chemical Physics Letters 462 (2008) 96-99.
43. S.S. Aplesnin, G.A. Petrakovskii, L.I. Ryabinkina, G.M. Abramova, N.I. Kiselev, O.B. Romanova 129 (2004) 195-197.
44. Felicia Iacomi, Aurelia Vasile, E.K. Polychroniadis, Materials Science and Engineering B 101 (2003) 275-278.
45. Felicia Iacomi, M. Vasilescu, S. Simon, Surface Science 600 (2006) 4323-4327.

Chapter III

INSTRUMENTATION AND CHARACTERIZATION TECHNIQUES

This chapter presents a brief description of various instrumentation and characterization techniques, which are used in the present work to characterize the synthesized undoped and cu-doped MnS nanocrystals.

3.1 Introduction

In the present work, different methods have been employed to characterize the synthesized undoped and doped MnS Nanocrystals. The powder X-ray diffraction (XRD) is used for confirmation of phase formation [1]. Morphology and topography of the materials are investigated through scanning electron microscope (SEM) and the transmission electron microscope (TEM). The Fourier transform infrared spectroscopy (FT-IR) technique is used to characterize the stretching and bending vibrational modes of different functional groups exist in the materials. The optical behavior and properties of the prepared samples are studied using UV-Visible spectroscopy and Photoluminescence spectroscopy (PL). UV-Visible absorption spectroscopy measures the percentage of radiation that is absorbed at each wavelength. PL technique involves measuring the energy distribution of emitted photons after optical excitation.

UV-Visible and Photoluminescence are used for studying the energy band gap values of materials and also to study the absorption and emission property of the materials. Actually the LCR meter (a pallet based system) is used for calculating the electrical properties of material [2]. This technique has been extensively exploited to understand the dielectric, impedance, electric modulus and ac electrical relaxation process in the material under the application of an ac electric field. The vibrating sample magnetometer (VSM) is used to measure the magnetic behavior of materials thereby extracting information about the paired/unpaired electron states in the material [3].

3.2 Structural Characterization of Nanocrystals
3.2.1 X-ray Diffraction (XRD)

In 1915, Braggs were awarded the Nobel Prize in physics for determining crystal structures beginning with NaCl, ZnS and diamond. Although Bragg's law was used to explain the interference pattern of X-rays scattered by crystals, diffraction has been developed to study the structure of all states of matter with a beam, its wavelength similar to the distance between the atomic or molecular structures of interest [4]. In 1919, A.W. Hull gave a paper titled, "A New Method of Chemical Analysis". It says that every crystalline substance gives a pattern; the same substance always gives the same pattern; and in a mixture of substances each produces its pattern independently of the others [5]. The X-ray diffraction pattern of a pure substance is, therefore, like a fingerprint of the substance. The powder diffraction method is used for characterization and identification of polycrystalline phases.

3.2.1.1 Bragg's Law

The law was first formulated by the English physicist Lawrence Bragg. Bragg's diffraction scheme is shown in Fig. 3.1. The rays of incident beam are always in phase and parallel up to the point at which the top beam strikes the top layer at atom A.

The second beam continues to the next layer where it is scattered by atom B. The second beam must travel the additional distance BC + BD if the two beams are to continue travelling adjacent and parallel [6].

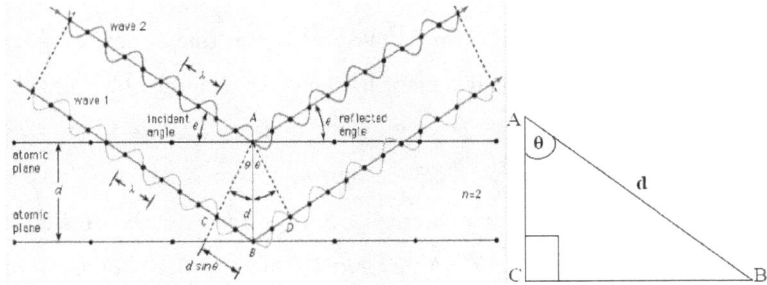

Fig. 3.1: Bragg's Diffraction Scheme

This additional distance must be an integral (n) multiple of the wavelength (λ) for the phases of the two beams to be the same, i.e.

$$n\lambda = BC + BD \quad (3.1)$$

Recognizing d as the hypotenuse of the right triangle ABC, we can use trigonometry to relate d and θ to the distance (BC + BD). Thus

$$BC = d \sin \theta \quad (3.2)$$

Since BC = BD then,

$$n\lambda = 2 \, BC \quad (3.3)$$

$$n\lambda = 2 \, d \sin \theta \quad (3.4)$$

where, θ is the angle of incidence of the X-ray; n is an integer; λ is the wavelength and d is the spacing between atom layers. The equation 3.4 is known as Bragg's Law [7]. Waves reflected through an angle corresponding to n=1 are said to be in the first order of reflection; the angle corresponding to n=2 is the second order, and so on. For any other angle (corresponding to fractional n) the reflected waves will be out of phase and will result in destructive interference. Bragg's law is useful for measuring wavelengths and for determining the lattice spacing of crystals.

3.2.1.2 X-ray Powder Diffractometer

In the powder, there are thousands of grains that have random orientations. It expects most of the different atomic planes to lie parallel to the surface in some of the grains. Thus, by scanning through an angle θ of incident X-ray beams form 0 to 90°, expect to find all angles where diffraction has occurred, and each of these angles would be linked with a different atomic spacing. The instrument used to do this is an x-ray powder diffractometer (Fig. 3.2). It consists of an X-ray tube able to producing a beam of monochromatic X-rays that can be rotated to produce angles from 0 to 90°. A powdered sample is placed on a sample stage and it can be irradiated by the X-ray tube. To detect the diffracted X-rays, an electronic detector is located on the other side of the sample from the X-ray tube, and it is allowed to rotate to produce angles from 0 to 90°. The instrument used to rotate both the X-ray tube and the detector is called a goniometer [8]. The goniometer keeps track of the angle θ, and sends this information to the computer, while the detector records the rate of X-rays coming out the other side of the sample (in units of counts/sec) and sends this information to the computer. The X-ray intensity can be plotted against the angle θ (reported as 2θ) to produce a chart.

The angle 2θ for each diffraction peak can then be converted to d-spacing, using the Bragg equation. One can then work out the crystal structure and correlate each of the diffraction peaks with a different atomic plane in terms of the Miller Index for that plane (hkl). A group known as the Joint Committee on Powder Diffraction Standards (JCPDS) has collected data such as this on thousands of crystalline substances. This data can be obtained as the JCPDS Powder Diffraction File. Since every compound with the same crystal structure will produce the same powder diffraction pattern, the pattern serves as kind of a "fingerprint" for the substance, and thus comparing an unknown sample to those in the Powder Diffraction file enables easy identification of the unknown [9].

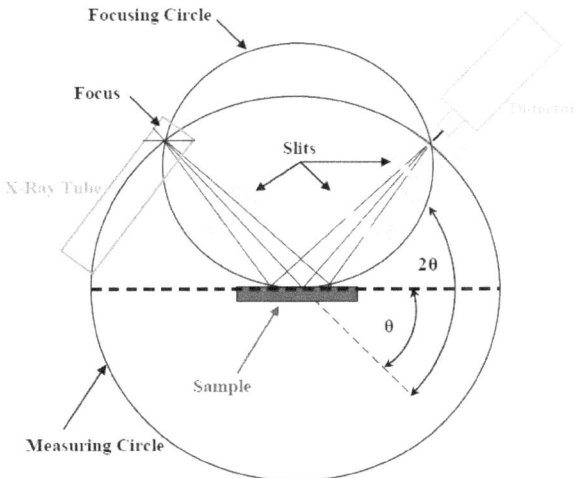

Fig. 3.2: Basic setup of X-Ray Diffractometer

The XRD equipment used in this study was a SHIMADZU XRD-6000 Powder Diffractometer with Cu Kα radiation at the Bragg angle ranging from 20° to 80°. The wavelength of the X-ray source employed is 1.54056 Å.

3.2.1.3 XRD Pattern Analysis

X-ray diffraction is a powerful tool for materials characterization as well as for detailed structural clarification. X-ray patterns are used to establish the atomic arrangements of the materials because of the fact that the lattice parameter, d (spacing between different planes) is of the order of x-ray wavelength. Further, x-ray diffraction method can be used to distinguish crystalline materials from nanocrystalline (amorphous) materials. The structure identification is made from the x-ray diffraction pattern analysis and comparing it with the internationally recognized database containing the reference pattern (JCPDS).

From x-ray diffraction pattern, the following information can be obtained:

- To judge formation of a particular material system.

- Unit cell structure, lattice parameters, miller indices, and types of phases.
- Estimation of crystalline/amorphous content in the sample.
- Evaluation of the average crystalline size from the width of the peak in a particular phase pattern.
- Analysis of structural distortion arising as a result of variation in d-spacing caused by the strain and thermal distortion.

3.2.1.4 Estimation of Crystallite size from XRD

The studies on crystalline materials require an accurate determination of crystallite size as well as the microstrain induced in the material. Based on XRD principles, numerous approaches such as the Scherrer equation, integral breadth analysis, single line approximation, Hall Williamson method, Rietveld refinement method have been developed for the crystallite size determination [10, 11]. An accurate estimation of crystallite size becomes essential when such materials are shaped with their crystallite size of the order of less than 100 nm. The XRD technique is simpler and easier approach for the determination of crystallite size of the samples [12].

The crystallite size, the lattice strain and instrumental effects accounts for the total broadening of XRD peaks. When the crystallites of a material are smaller (less than 100nm) they have too small a number of parallel diffraction planes and eventually they produce broadened diffraction peaks instead of sharp peaks. Lastly, instrumental factors such as unresolved α_1 and α_2 peaks for imperfect focusing also lead to line broadening. The observed peak broadening β_0 may be represented as

$$\beta_0 = \beta_i + \beta_r \qquad (3.5)$$

Here β_0 is the observed peak broadening in radians; β_i is broadening due instrumental factors, in radians and β_r is broadening due crystallite size and lattice strain.

However, when the diffraction peaks show a mixed behavior (partly Cauchy and partly Gaussian) for their profiles, the following equation holds well [13, 14],

$$\beta_r = \left[(\beta_0 - \beta_i)(\beta_0^2 - \beta_i^2)^{1/2} \right]^{1/2} \qquad (3.6)$$

Now according to Scherrer equation, the broadening due to small crystallite size can be written as [15],

$$\beta_r = k\lambda / S_c \cos\theta \qquad (3.7)$$

Here β_r is broadening only due to small crystallite size; k is constant whose value depends on particle shape and usually taken as 1.0; S_c is the small crystallite size in nanometers; θ is the Bragg's angle; λ is the wavelength of x-ray beam in nm.

According to Wilson the broadening due to lattice strain may be expressed by the relation [15],

$$\beta_s = 4e \tan\theta \qquad (3.8)$$

Here β_s is peak broadening due to lattice strain; e is the strain distribution within the material.

Using the equation 3.7 and equation 3.8, the total peak broadening

$$\beta_r = k\lambda / S_c \cos\theta + 4e \tan\theta \qquad (3.9)$$

Multiplying $\cos\theta$ on both sides of the above equation, then one can get [15]

$$\beta_r \cos\theta = k\lambda / S_c + 4e \sin\theta \qquad (3.10)$$

The above equation is Hall-Williamson equation. It represents the Uniform Deformation Model (UDM), where the strain was assumed to be uniform in all crystallographic directions, thus considering the isotropic nature of the crystal, where all the material properties are independent of the direction along which they are measured.

The term ($\beta_r \cos\theta$) is plotted with respect to ($4\sin\theta$) for the preferred orientation peaks of as-prepared samples. Accordingly, the slope and y-intersect of the fitted line represent strain and particle size, respectively. W.H method is the accurate method than Scherer method for crystallite size estimation of powder samples in view of the ability of the former approach [16].

3.2.1.5 Lattice Parameter

The lattice parameters a and c of as-prepared sample is calculated using the following formula for the hexagonal crystal structure [17],

$$d_{hkl} = \frac{1}{\sqrt{\frac{4}{3}\left(\frac{h^2+hk+k^2}{a^2}\right)+\frac{l^2}{c^2}}} \qquad (3.11)$$

where $d_{hkl} = \frac{n\lambda}{2\sin\theta}$; h, k, and l are all integers; (h k l) is the lattice plane index; a and c are lattice constants, d_{hkl} is distance between two consecutive planes (n = 1) with lattice plane index; λ is the wavelength of x-ray beam in nm; θ is the Bragg's angle.

The lattice parameter of zinc-blende or sphalerite structured nanomaterials can be calculated using the following formula [18],

$$a = d_{hkl}\left(h^2+k^2+l^2\right)^{1/2} \qquad (3.12)$$

where, d_{hkl} Bragg spacing between the reflecting faces, which is the width of the unit cells between the planes; (h k l) is the lattice plane index; a is the lattice constant.

3.2.1.6 Specific Surface Area

The surface states will play an important role in the nanoparticles, due to their large surface to volume ratio with a decrease in particle size. The Specific Surface Area (SSA) is derived scientific value that can be used to determine the type and properties of a material.

It has a particular importance in case of adsorption, heterogeneous catalysis and reactions on surfaces. The specific surface area and surface to volume ratio increase significantly as the size of materials decreases. Mathematically, SSA can be calculated using formulas (3.13 & 3.14) and both formulas yield same result [19].

$$SSA = \frac{SA_{part}}{V_{part} * density} \quad (3.13)$$

$$SSA = \frac{6*10^3}{D_p \rho} \quad (3.14)$$

Where SSA is the specific surface area, V_{part} is particle volume and SA_{part} is Surface Area, D_p is the crystallite size (spherical shaped), and ρ is the density of nanomaterials.

3.2.1.7 Dislocation Density

The dislocation density is the length of dislocation lines per unit volume of the crystal. A dislocation is a crystallographic defect, or irregularity, within a crystal structure. The presence of dislocation strongly influences various properties of materials. Mathematically, it is a type of topological defect. It increases with plastic deformation; a mechanism for the creation of dislocations must be activated in the material. Dislocations are formed by three mechanisms i.e. homogeneous nucleation, grain boundary initiation and dispersed phases. The movement of a dislocation is impeded by other dislocations present in the sample. Thus, a larger dislocation density implies a larger hardness. Chen and Hendrickson measured and determined dislocation density and hardness of several silver crystals. They found that crystals with larger dislocation density were harder. It has been shown that the dislocation density increases while the grain size decreases with increasing strain and ultimately these parameters reach saturation values [20].

The x-ray line profile analysis has been adopted to determine dislocation density. The dislocation density (δ) in the sample has been determined using expressions (3.15 & 3.16) and results from both the formulas are approximately same [21].

$$\delta = \frac{15\beta \cos\theta}{4\alpha D} \quad (3.15)$$

$$\delta = \frac{1}{D^2} \quad (3.16)$$

where, δ is dislocation density, β is diffraction broadening (radian), θ is diffraction angle (degree), α is lattice constant (nm) and D-particle size (nm).

3.2.2 Electron Microscopy

In order to determine the morphology and the size of nanoparticles, an electron microscope is used. An electron microscope is a type of microscope that uses a particle beam of electrons to illuminate a specimen and create a highly-magnified image. The better resolution and magnification of the electron microscope is because the wavelength of an electron is much smaller than that of a photon of visible light. The electron microscope uses electrostatic and electromagnetic lenses in forming the image by controlling the electron beam to focus it at a specific plane relative to the specimen. The most commonly used electron microscopes are: Transmission Electron Microscope (TEM), Scanning Electron Microscope (SEM), Reflection Electron Microscope (REM), and Scanning Transmission Electron Microscope (STEM). The electron microscopes used in this work are the TEM and SEM.

3.2.2.1 Transmission Electron Microscopy (TEM)

The transmission electron microscope (TEM) uses a high voltage electron beam to create an image. In Fig. 3.3, the basic principle of a TEM is shown schematically.

The electrons are generated in the electron source, which consists of a field emission gun (FEG) or thermionic gun. FEG creates high electric field typically at +100 keV (40 to 400 keV) with respect to the cathode, which generates the electrons via the quantum mechanical tunnelling effect from the cathode to the vacuum. In thermionic gun, the cathode is heated up to very high temperatures. If the energy gain for electrons is greater than the work function of the cathode material the electron will be released from the cathode. After the electrons have been generated, an electric field with a voltage U will be applied in order to accelerate them to achieve the desired wavelength. The velocity of the electrons is near to the speed of light, the calculation of the wavelength should be relativistic. Combination of the Eq. (3.17) and (3.18) gives the wavelength of the electrons, which depends on the accelerating voltage U [22].

$$E = \sqrt{(pc)^2 + m_0^2 c^4} \qquad (3.17)$$

$$\lambda = h/p \qquad (3.18)$$

$$\lambda = \frac{h}{\left[2m_0 eU\left(1 + \frac{eU}{2m_0 c^2}\right)\right]^{1/2}} \qquad (3.19)$$

where e is the electron charge and $m_0 c^2$ is the rest energy of the electron (0.5 MeV). For an accelerating voltage of 100 and 200 kV the wavelength of the electron is 3.9 and 2.7 pm, respectively.

In addition the several condenser lenses are used to direct and vary the electron beam towards the thin specimen. The several projective lenses are used to direct the transmitted electrons to the viewing screen or the CCD camera. An objective aperture is used, between the objective and projective lenses, in order to select certain electron beams for different imaging modes.

One can change the magnification for example by varying the strength of the magnetic fields in above lenses. The image detected by the CCD may be displayed on a monitor or computer.

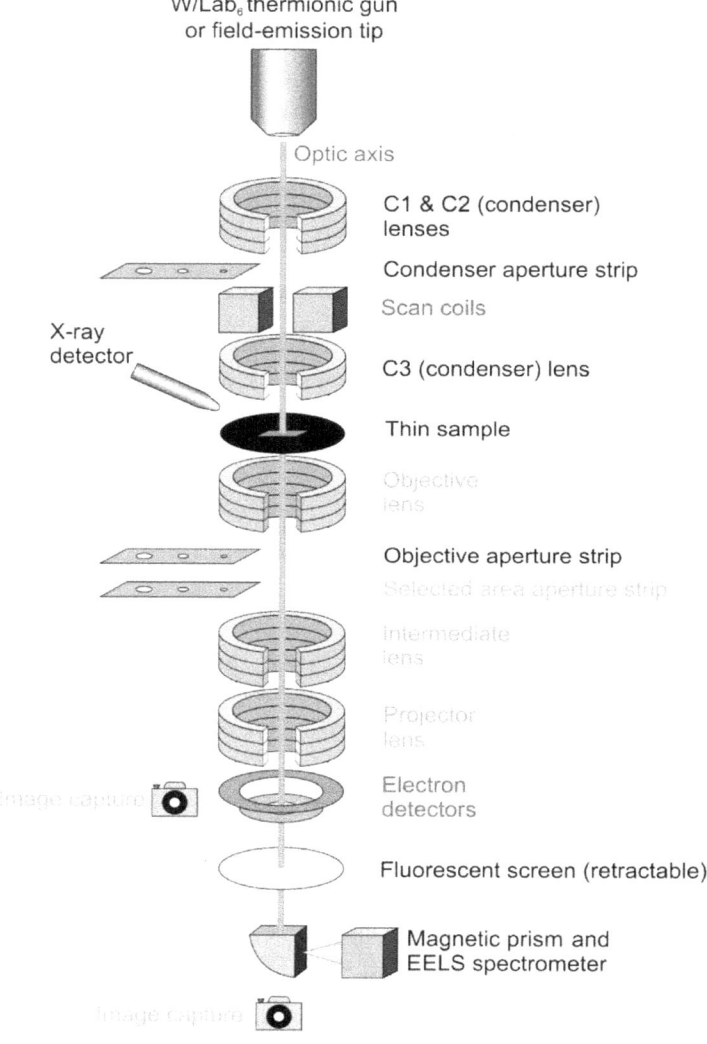

Fig. 3.3: Basic principle of a TEM

3.2.2.2 Scanning Electron Microscope (SEM)

A scanning electron microscope (SEM) produces images of a sample by scanning it with a focused beam of electrons. The electrons interact with atoms, producing various signals that contain information about the sample's surface topography and composition. The electron beam position is combined with the detected signal to produce an image. SEM can achieve resolution better than 1 nanometer. The specimens can be observed in a vacuum at wide range of cryogenic or elevated temperatures. The most common SEM mode is detection of secondary electrons emitted by atoms excited by the electron beam. Number of secondary electrons depends on angle at which beam meets surface of specimen, i.e. on specimen topography. By scanning the sample and collecting the secondary electrons with special detector, an image displaying the topography of the surface is created [23]. The schematic representation of a SEM is shown in Fig. 3.4.

In a typical SEM, an electron beam is thermionically emitted from an electron gun fitted with a tungsten filament cathode. The electron beam, which usually has an energy ranging from 0.2 keV to 40 keV, is focused by one or two condenser lenses to a spot about 0.4 nm to 5 nm in diameter. The beam passes through pairs of deflector plates in the electron column, typically in the final lens, which deflect the beam in the x and y axes so that it scans in a raster fashion over a rectangular area of the sample surface. When the primary electron beam interacts with the sample, the electrons lose energy by repeated random scattering and absorption within the interaction volume, which extends from less than 100 nm to approximately 5 µm into the surface. The size of the interaction volume depends on the electron's landing energy, the atomic number of the specimen and the specimen's density.

The energy exchange between the electron beam and the sample results in the reflection of high-energy electrons by elastic scattering, emission of secondary electrons by inelastic scattering and the emission of electromagnetic radiation, each of which can be detected by dedicated detectors. The beam current absorbed by the specimen can also be detected and used to create images of the distribution of specimen current. Electronic amplifiers are used to amplify the signals, which are displayed as variations in brightness on a computer screen. Each pixel of computer video memory is synchronized with the position of the beam on the specimen in the microscope, and the resulting image is therefore a distribution map of the intensity of the signal being emitted from the scanned area of the specimen.

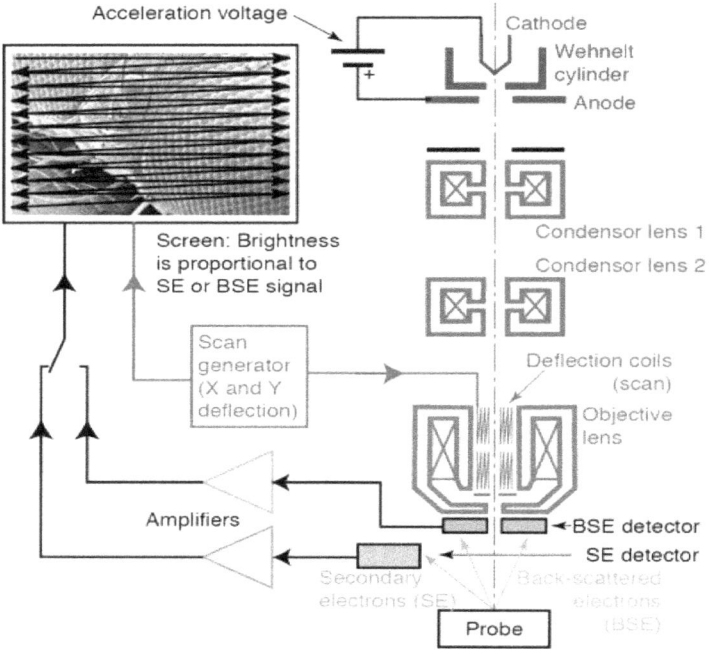

Fig. 3.4: Schematic representation of a SEM

The Transmission Electron Microscopy (TEM) and Scanning Electron Microscope (SEM) systems employed in this work are a HITACHI H-7500 and a SEM JEOL- 6390, respectively.

3.3 Optical Characterization of Nanocrystals

Optical properties are normally characterized by using spectroscopic techniques including UV-visible and photoluminescence spectroscopy, which both yield information about the electronic structure of nanoparticles. It can be used for calculation of bang gap energy and determination of defects in the crystals. The optical techniques such as Raman and FTIR spectroscopy provide information about the crystal structure such as photon or vibrational frequencies and crystal phases [24]. The synthesized nanocrystal is characterized with, Fourier transform infrared, UV-Visible absorption and photoluminescence spectroscopic techniques.

3.3.1 Fourier Transform Infrared Spectroscopy

Fourier Transform Infrared spectroscopy (FTIR) provides information about the chemical bonding in a material. The aim of the basic infrared experiment is to determine changes in the intensity of a beam of infrared radiation as a function of wavelength or frequency after it interacts with the sample. The ratio of the intensity before and after the light interacts with the sample is determined. The plot of this ratio versus frequency is the infrared spectrum. The infrared spectrum originates from the vibrational motion of the molecule. The vibrational frequencies are a kind of fingerprint of the compounds and are used to characterize it. The design of optical pathway produces a pattern called interferogram, which is a time domain spectrum containing intensity vs time. Fourier transform separates the individual absorption frequencies from the interferogram and better signal to noise ratio is plotted, hence it has more speed and sensitivity.

The FT-IR uses an interferometer to process the energy sent to the sample, which contain a beam splitter, as shown in Fig. 3.5, which divides the incident beam to two perpendicular component one is undeflected and the other is at 90° angle. The beam oriented at 90° goes to a fixed mirror and is returned to the beam splitter.

Motion of mirror causes the path length of the second beam to vary. Combined beam containing the interference is called interferogram, which contain all radiative energy coming from the source and has wide range of wavelengths. Beam splitter orients the interferogram towards the sample; that absorbs all the wavelengths which are present in the infrared spectrum. The modified interferogram signal that reaches the detector contains information about amount of energy that is absorbed at each frequency [25].

The energy corresponding to the transitions between molecular vibrational states is generally 1-10 kilocalories/mole which corresponds to the infrared portion of the electromagnetic spectrum.

Difference in Energy states = Energy of Light absorbed

$$E_1 - E_0 = \frac{hc}{\lambda} \qquad (3.20)$$

where, h is the Planks constant; c is speed of light; λ is the wavelength of light.

In order to record the FT-IR spectrum of a sample, it should be placed in cell that is made out of a material which is inert to the infrared frequencies. Usually, KBr or NaCl is used for making the cells, but NaCl begins to absorb at 650cm^{-1} and bands with frequencies less than this value will not be observed [26]. For recording the FT-IR spectrum, solid samples mixed with KBr are made to pellet by pressing at high pressure.

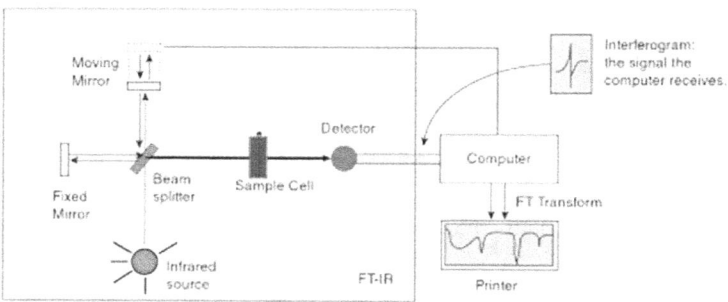

Fig. 3.5: Schematic of Fourier Transform infrared spectrometer

In this study, FTIR spectra have been recorded using a SHIMADZU FTIR-8400S spectrometer in the wave number range of 400-4000 cm^{-1}.

3.3.2 Ultraviolet–Visible Spectroscopy

Optical spectroscopy has been widely used for the characterization of nanomaterials and the techniques can be generally categorized into two groups: absorption and emission spectroscopy. The electronic structures of atoms, ions, molecules or crystals through exciting electrons from the ground to excited states (absorption) and relaxing from the excited to ground states (emission). Ultraviolet and visible absorption spectroscopy is the measurement of light when it is passed through a sample.

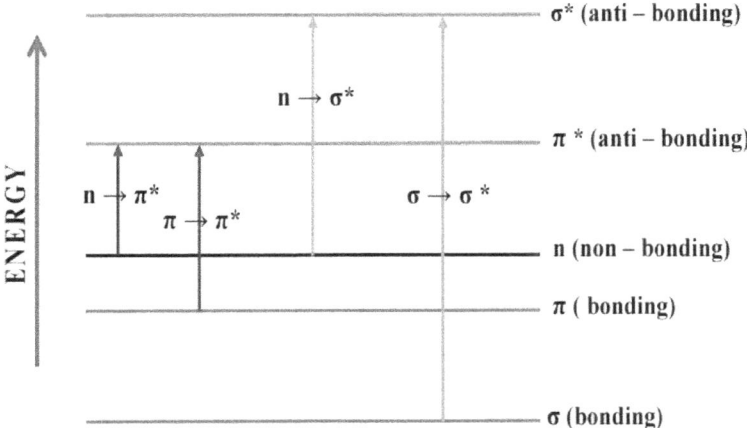

Fig. 3.6: The electronic energy levels and transitions

The principle of UV-Vis spectroscopy is based on the ability of molecule to absorb ultraviolet and visible light. The absorption of light corresponds to the excitation of outer electrons in the molecule. When a molecule absorbs energy and the outer electrons in the molecule excited from the Highest Occupied Molecular Orbital (HOMO) to Lowest Unoccupied Molecule Orbital (LUMO). The occupied molecular orbitals with lowest energy are known the σ orbitals, at slightly higher energy are called π orbitals and at still higher energy are known non-bonding orbitals (unshared pair electrons).

The π* and σ* are called the highest energy state. The Fig. 3.6 shows the electronic energy levels and transitions [27]. Ultraviolet and visible spectroscopy uses electromagnetic radiations between 200 nm to 800 nm and is divided into the ultraviolet (UV, 200-400 nm) and visible (VIS, 400-800 nm) regions which covers the UV-visible region of energy for the electromagnetic spectrum 1.5 - 6.2 eV.

3.3.2.1 Beer–Lambert Law

The larger the number of molecules that absorb light of a given wavelength, greater the light absorption and higher the peak intensity in absorption spectrum. If there are only a few molecules that absorb radiation, the total absorption of energy is less and thus lower intensity peak is observed. This makes the basis of Beer-Lambert Law which states that the fraction of incident radiation absorbed is proportional to the number of absorbing molecules in its path. When the beam of monochromatic light passes through a solution, the amount of light absorbed or transmitted is an exponential function of the molecular concentration of the solute and also a function of length of the path of radiation through the sample [28]. Therefore,

$$Log(I_0/I) = \varepsilon.c.l \tag{3.21}$$

where, I_0 is an intensity of the incident light; I is an intensity of light transmitted through the sample solution; c is the concentration of the sample and expressed as mol L^{-1}; l is the path length of the sample and expressed in units cm; ε = molar absorbtivity and expressed in units L mol^{-1} cm^{-1}.

The ratio I/I_0 is known as transmittance T and the logarithm of the inverse ratio I_0/I is known as the absorbance A.

$$\log(I_0/I) = A = \varepsilon cl \tag{3.22}$$

For presenting the absorption characteristics of a spectrum, the positions of peaks are reported as $\hat{\lambda}_{max}$ (in nm) values.

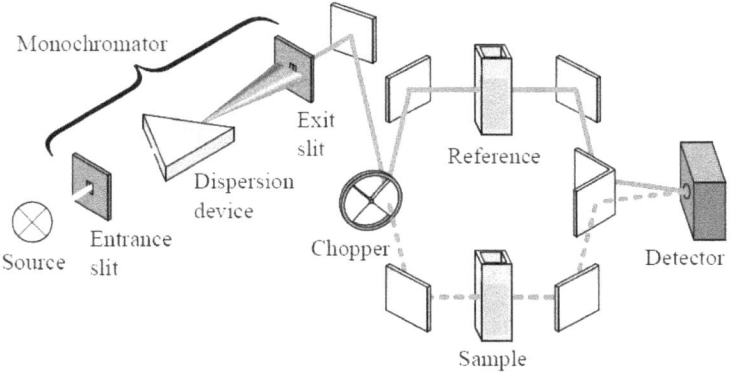

Fig. 3.7: Schematic of a dual-beam spectrophotometer

UV spectroscopy obeys the Beer-Lambert law [29]. Fig. 3.7 shows a schematic of a dual-beam spectrophotometer. The most of the spectrophotometers are double beam spectrophotometers. The light source provides the visible and near ultraviolet radiation covering the 200 - 800 nm. The output from the light source is focused onto the diffraction grating which splits the incoming light into its component colors of different wavelengths, like a prism. In this configuration, a chopper is placed in the optical path, near the light source. The chopper switches the light path between a reference optical path and a sample optical path to the detector. It rotates at a speed such that the alternate measurements of blank and sample occur several times per second, thus correcting for medium and long term changes in lamp intensity. In this work, the absorbance spectra have been recorded by using a spectro- photometer JASCO V-570.

3.3.3 Photoluminescence (PL) Spectroscopy

In semiconductors, luminescence is a radiative recombination process involving electrons and holes. An electron-hole pair or exciton are created due to an excitation, the carriers then recombine and a photon is emitted from the system.

This excitation can be thermal, electrical and optical. If the excitation is thermal, the process is known as thermo-luminescence, if it is electrical then the process is known as electroluminescence and if it is optical then the process is known as photoluminescence (PL) [30]. This section will describe the mechanism for photoluminescence and its experimental setup. An electron, which is excited above the CB, and a hole, which is excited below the VB, can gain a kinetic energy that is higher than the thermal energy $k_B T$. Both tend to lose this energy very rapidly by releasing acoustic and longitudinal phonons. This process is called carrier relaxation [31].

Fig. 3.8 presents the photoluminescence process in a direct band gap semiconductor. PL spectroscopy is a powerful technique to probe the optical and electronic properties of semiconductor materials. PL spectroscopy can be used to estimate the band gap energy, defect/impurity detection and to look at recombination mechanisms. By probing the energy levels, PL presents valuable information about the composition of the semiconductor and the effect of confinement. Also, the line width of the PL emission can be used to understand the role of disorder and lattice vibrations on the radiative recombination process [32].

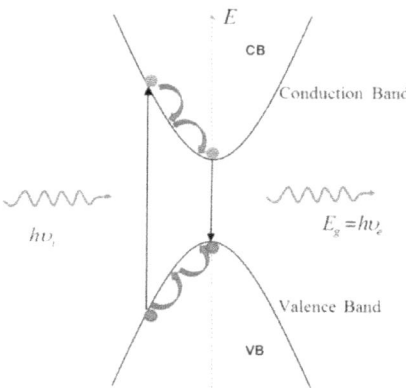

Fig. 3.8: Photoluminescence process in a direct band gap material

3.3.3.1 PL Experimental Setup

A typical PL setup is shown in Fig. 3.9. The simplicity of the experimental setup means that the experiment can be performed at low temperatures. The most common source for excitation in PL experiments is lasers. Lasers produce intense, monochromatic light making them an excellent excitation source. When the laser beam is incident on the sample, photoluminescence occurs and light is emitted from the sample at wavelengths dependent on the sample composition.

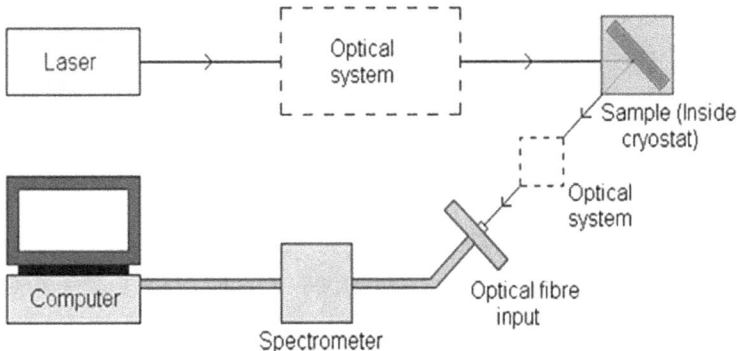

Fig. 3.9: Typical setup for the measurement of PL

The emitted light is directed into a fibre optic cable and then into a spectrometer. A filter may be placed in front of the fibre input to remove any incident laser light. Inside the spectrometer, a diffraction grating diffracts different wavelengths in different directions towards an array of photo-detectors that measure the intensity of each wavelength component. The digital information is interpreted by the computer, which can display a PL spectrum. The measurement of photoluminescence from semiconductor materials has become an important characterization method and is widely used to provide information on, e.g. carrier doping levels, alloy compositions, film structures, band gap and edge effects, etc. in applications ranging from scientific research, process monitoring and device characterization.

In this study, room temperature PL spectra have been recorded using a HORIBA JOBIN YVON-Fluorolog at an excitation wavelength of 350 nm.

3.4 Electrical Characterization of Nanocrystals

Impedance spectroscopy is a powerful method for characterizing many of the electrical properties of solid materials and their interfaces. The instrument can directly measure 14 impedance parameters $\left(|Z|, |Y|, \theta, C_s, C_p, D, L_s, L_p, Q, R_s, R_p, G, X, \& B\right)$. Out of these 14 parameters few parameters can be displayed simultaneously. The electrical properties such as impedance, dielectric constant, loss, ac conductivity and electric modulus are presented in order to correlate the conduction mechanism and relaxation phenomena [33].

In this present work, the electrical studies were carried out by using Hioki 3532-50 LCR HiTester in the frequency range of 50Hz to 5 MHz.

3.4.1 LCR Meter

The impedance study can be used to obtain the grain boundary effects, ionic transport, conductivity, double layer formation at the electrode/electrolyte interface, etc. In recent years, Impedance spectroscopy (IS) is used for characterizing the electrical properties of the nanostructured materials and their interfaces with electronically conducting electrodes [34]. The complex impedance (Z^*), complex admittance (Y^*), complex dielectric permittivity (ξ^*), complex modulus (M^*) and AC conductivity (σ) are calculated from the relation,

Complex impedance $Z^* = Z' - jZ''$ (3.23)

Complex admittance $Y^* = Y' + jY''$ (3.24)

Complex permittivity $\varepsilon^* = \varepsilon' - j\varepsilon'' = 1/j\omega C_0 Z^*$ (3.25)

Complex modulus $M^* = M' + jM'' = j\omega C_0 Z^*$ (3.26)

Where C_0 is the vacuum capacitance and ω=2πf is the angular frequency.

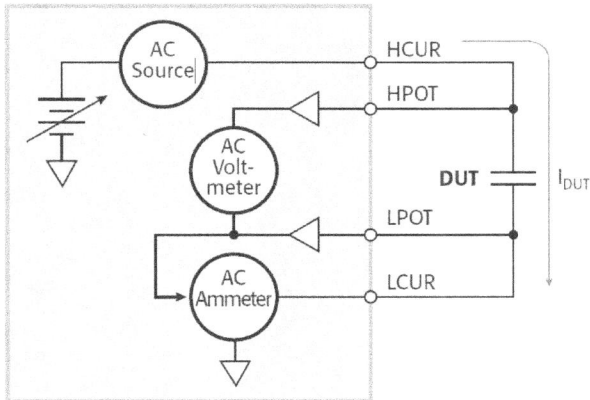

Fig. 3.10: AC impedance meter

The operation of LCR meter (Fig. 3.10) is relatively simple. It measures AC impedance by supplying an AC voltage out of the high current terminal (HCUR). The current through the device is measured by the low current terminal (LCUR), and the voltage across the device is measured by the high and low potential terminals (HPOT and LPOT). The voltage and current are measured in a phase-locked manner that precisely identifies the phase angle between them. By knowing the amplitude and phase angle, it's possible to calculate any desired AC impedance parameter.

The AC impedance parameters mean measuring the amplitude of the impedance, which is designated in the graph in Fig. 3.11 as 'Z'. It also requires measuring the phase angle between the current and the voltage, designated as 'θ'. It is also possible to convert that mathematically to a rectangular form, which would result in the R+jX designation. R represents the real or in-phase impedance vector, while jX represents the imaginary or 90° out of phase impedance vector. It is even possible to convert polar and rectangular forms mathematically into actual capacitance and resistance values [35].

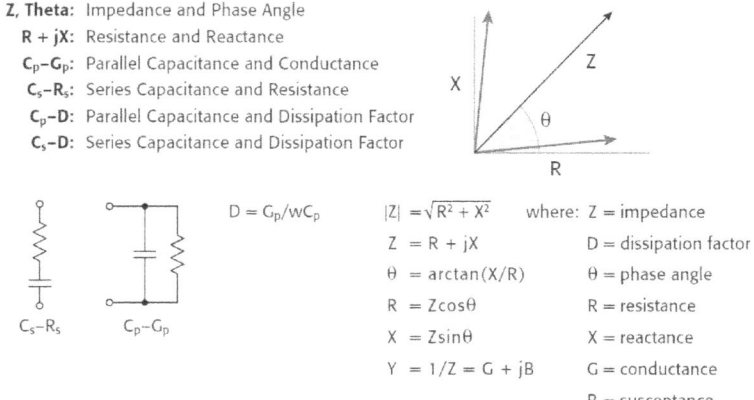

Fig. 3.11: Basic AC impedance parameters

The method of four-terminal measurement is used to reduce the contact resistance and to minimize any stray capacitance and residual inductance associated with the test leads or the test fixture at high frequencies. Four coaxial cables are used as leads from the BNC connector on the LCR meter to the sample holder as shown in Fig. 3.12. These coaxial cables are connected, at one end, to the high current, to the high potential, to the low potential, and to the low current ports of the LCR meter. The other ends of these cables are connected to four terminals on the sample holder.

The four outer conductors of these cables are short-circuited at the end near the sample holder, they serve as the return path for the measurement current. The same current flows through both the outer shield conductors and the center conductors so that no external magnetic fields are generated around the conductors. Thus, the test leads do not contribute any additional measurement error due to the self or the mutual inductance between the individual leads.

Since the measurement circuit has residual inductances, inherent stray capacitances, and resistances, the measured values may be unacceptably influenced depending on the measurement range and the magnitude of the residual parameters [36].

In the present investigation the electrical study was carried out by the pressed pellets of samples of known dimension on HIOKI 3532-50 HITESTER precession LCR meter using specially designed sample holder, within the frequency range from 50Hz to 5MHz. The samples are prepared and mounted between the copper platforms and electrodes. In order to ensure good electrical contract between the crystal and silver paint does not spread to the sides of the crystal. The capacitance and the dissipation factor of the parallel plate capacitor formed by the copper plate and electrode having the sample as a dielectric medium are measured.

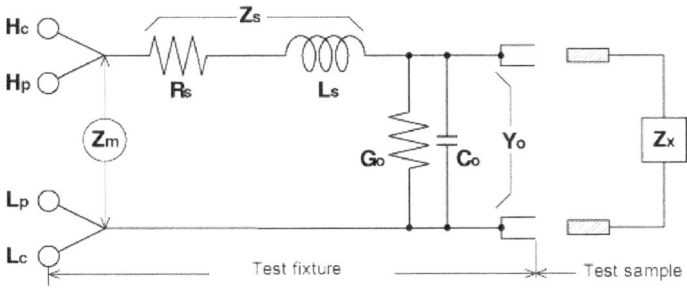

Zx : true value
Ls : residual inductance
Co : floating capacitance value
Zs : short circuit residual component
Yo: open circuit residual component

Rs : residual resistance
Go : residual conductance
Zm : measured value

Fig. 3.12: Schematic diagram of four probe LCR meter

3.4.2 Elementary Analysis of Impedance Spectra

AC electrical response of various combinations of R and C elements and the response of solid electrolyte-electrode system to the applied electric field are described. The non-linear least square fit technique to extract the electrical parameters from the complex impedance representation is illustrated. There are two common AC impedance models: the parallel model and the series model [37]. In the parallel model, results are expressed as the parallel capacitance (C_p) and the parallel conductance (G_p). In the series model, results are expressed as the series capacitance (C_s) and the series resistance (R_s). The dissipation factor (D), the ratio of the real impedance to the imaginary impedance, is another common parameter that is mathematically derived. The admittance and permittivity are parallel functions characteristic at low frequencies whereas the impedance and modulus are series functions at high frequencies.

In series: The complex impedance for the series combination of resistance R_s and capacitance C_s is given by

$$Z^* = R_s - 1/\omega C_s \qquad (3.27)$$

$$Z^* = Z' - jZ'' \qquad (3.28)$$

Fig. 3.13 (a) shows the resultant impedance for R & C in series is represented by a vertical line parallel to the imaginary axis intersecting the real axis at R. The corresponding admittance plot gives the semicircle intersecting the real axis at the origin and at a point 1/R as shown in Fig 3.14 (c).

In parallel: the complex impedance for the parallel combination of R_p and C_p is given by

$$1/Z^* = 1/R_p + j\omega C_p \qquad (3.29)$$

where, $Z^* = Z' - jZ''$

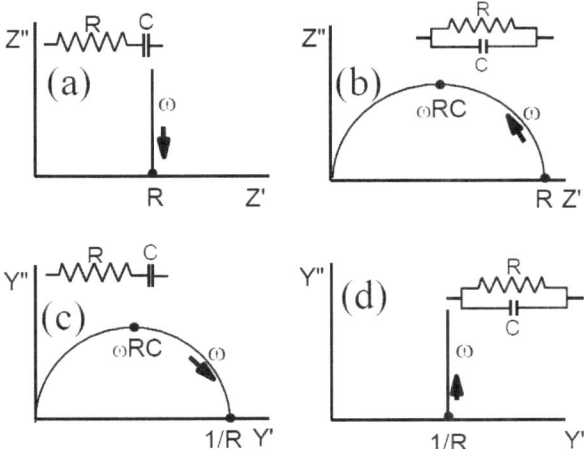

Fig. 3.13: (a)-(d) Complex impedance and admittance spectra for series and parallel combinations of R and C

The complex impedance for a parallel circuit represents a semicircle, as shown in the Fig. 3.13 (b), intersects the real axis at origin and as a point R. The Z" has the maximum value, R/2 at ωRC=1. The difference between the two intercepts gives the bulk resistance (R_b) of the material [38-40]. The impedance and admittance plane representations are commonly used for analyzing the response of solid electrolytes. Alternatively, the complex modulus and complex permittivity spectra are used for representing the response of the dielectric systems.

3.4.3 Dielectric Studies

In the presence of an electric field, the cloud of negatively charged electrons of an atom in dielectrics is distorted, making dipoles as shown in the left of the Fig. 3.14. As each dipole is characterized by its dipole moment thus they produce their own field, which interact with the external applied field [41, 42].

The process of relative displacement of the negative and positive charges of atoms or molecules, the orientation of existing dipoles toward the direction of the field, or the separation of mobile charge carriers at the interfaces of grain-boundaries, caused by an external electric field is referred to as an electric polarization [43, 44]. There are four types of polarization, as electronic, Ionic or atomic, dipolar and interfacial or space-charge polarization. Electric polarization associated with mobile and trapped charges is generally referred to as interfacial or space-charge polarization. This mainly occurs in amorphous or polycrystalline materials of traps and charge carriers (electrons, holes) [45].

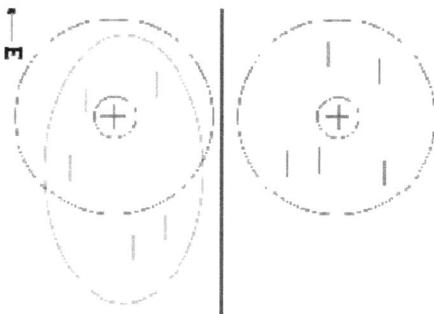

Fig. 3.14: Electric field interactions with an atom under the classical dielectric model

Permittivity is determined by the ability of a material to polarize in response to the field, and in that way reduce the total electric field inside the material. Thus, permittivity relates to the ability of material to transmit an electric field. The response of normal materials to external fields usually depends on the frequency of the field. This frequency dependence reflects the fact that the polarization of material does not respond instantaneously to an applied field. The response must always be causal which can be represented by a phase difference. For this reason permittivity is often treated as a complex function. The response of materials to alternating fields is characterized by a complex permittivity [46],

$$\varepsilon^* = \varepsilon' - j\varepsilon'' = |\varepsilon| e^{-j\delta} \qquad (3.30)$$

where ε' is the real part of the relative permittivity (i.e. the dielectric constant), which is related to the stored energy within the medium; and ε'' is the imaginary part of the relative permittivity, which is related to the dissipation (or loss) of energy within the medium. Equation (3.30) expresses the complex permittivity in two parts, as real and imaginary or as magnitude and phase.

3.4.4 Electric Modulus

Electrical dispersion phenomena in ionic conductors, which do not have permanent molecular dipoles, appears from the long range ionic diffusion process involving jumps over free energy barriers of varying heights. The dispersion connected to the long range ionic diffusion process is called conductivity relaxation, to discriminate it from the dispersion due to permanent dipole reorientation [47]. Ionic conductivity is a series process involving consecutive hops of an ion over the potential energy barriers in the direction of electric field. As the impedance are additive for a series process rather than admittance, for analyzing conductivity relaxation a dimensionless parameter called complex electric modulus, $M^*(\omega)$, which is the inverse of complex permittivity and is defined as [48, 49]:

$$M^*(\omega) = \frac{1}{\varepsilon^*(\omega)} = M' + jM'' = \frac{\varepsilon'}{\varepsilon'^2 + \varepsilon''^2} + j\frac{\varepsilon''}{\varepsilon'^2 + \varepsilon''^2} \quad (3.31)$$

The modulus function is defined as the Fourier transform of Kohlrausch-Williams-Watts relaxation function, $\phi(t)$. It is given by:

$$M'(\omega) = M_\infty \left[1 - \int_0^\infty \exp(-j\omega t) \frac{d(\phi(t))}{dt} dt \right] \quad (3.32)$$

where, $M_\infty = \dfrac{1}{\varepsilon_\infty}$, where, ε_∞ is the high frequency value of the real part of dielectric permittivity.

The $\phi(t)$, describes the evolution of electric field within the material for a macroscopic current decay after an electric field change and is of the form [50, 51]:

$$\phi(t) = \exp\left[-\left(\frac{t}{\tau_p}\right)^\beta\right] \qquad (3.33)$$

where, τ_p is the conductivity relaxation time and $0<\beta\leq 1$ is an exponent that measures the extent of non-exponential behavior and tend towards unity for Debye type relaxation. Thermally activated average conductivity relaxation time for the ion motion is given by:

$$\tau_p = \tau_0 \exp\left[\frac{E_h}{k_B T}\right] \qquad (3.34)$$

where, E_h is activation energy for hopping and k_B is the Boltzmann constant.

3.4.5 AC Conductivity

If a capacitor is charged by the AC voltage, there will be a loss current due to impedance by heat absorption. The AC conductivity (σ_{ac}) due to AC voltage is given by the relation [52, 53],

$$\sigma_{ac} = \frac{J}{\vec{E}} \qquad (3.35)$$

$\vec{E}$ is the electric field intensity vector and J is the current density. For a parallel plate capacitor the electric field intensity ($\vec{E}$) is the ratio of potential difference between the plates to the inter-plate distance of the capacitor, i.e.

$$\vec{E} = V/d \qquad (3.36)$$

Since the current density $J = \left(\frac{1}{A}\right)\frac{dQ}{dt}$, where Q is given by

$$Q = V\varepsilon A/d \qquad (3.37)$$

Therefore,

$$J = d\left(\frac{V\varepsilon/d}{dt}\right) = (\varepsilon/d)\frac{dV}{dt} \quad (3.38)$$

Hence, The AC conductivity can be written as

$$\sigma_{ac} = \frac{J}{\vec{E}} = \varepsilon j\omega \quad (3.39)$$

where, $\varepsilon = \varepsilon' - j\varepsilon''$ is a complex quantity.

$$\sigma_{ac} = (\varepsilon' - j\varepsilon'')j\omega = \varepsilon' j\omega + \omega\varepsilon'' \quad (3.40)$$

We know that the AC conductivity is a real quantity and the term containing j has to be neglected. Thus,

$$\sigma_{ac} = \omega\varepsilon'' \quad (3.41)$$

The two components ε' and ε'' of the complex dielectric constant (ε) will be frequency dependent and is given by [54]

$$\varepsilon'(\omega) = D_0 \cos\delta / \vec{E}_0 \quad (3.42)$$

$$\varepsilon''(\omega) = D_0 \sin\delta / \vec{E}_0 \quad (3.43)$$

Generally, sin δ is called the loss factor but when δ is small then sin δ = δ = tan δ. Since the displacement vector in a time varying field will not be in phase with $\vec{E}$ there will be a phase difference δ between them. From (3.42) and (3.43), we have

$$\tan\delta = \frac{\varepsilon''}{\varepsilon'} \quad (3.44)$$

Substituting the value of ε'' in σ_{ac}, we have

$$\sigma_{ac} = \omega\tan\delta\varepsilon' \quad (3.45)$$

where $\omega = 2\pi f$ and $\varepsilon' = \varepsilon_0\varepsilon_r$, where ε_r is the relative permittivity of the material and ε_0 is the permittivity of the free space. Then we get,

$$\sigma_{ac} = 2\pi f \tan\delta\varepsilon_0\varepsilon_r \quad (3.46)$$

The above equation is used to calculate the AC conductivity. It is to be noted that both tan δ and εr were available from dielectric measurement [55, 56]. The plots between ln(σac) and 1000/T were found to be very linear. The conductivity values can be fitted to the relation

$$\sigma_{ac} = \sigma_0 \exp[-E_a / kT] \qquad (3.47)$$

where E_a is the activation energy, k is the Boltmann constant, T is the absolute temperature and σ_0 is the pre-exponential factor. The above equation is called the Arrhenius equation.

$$E_a = -(slope)k \times 1000 \qquad (3.48)$$

The linear regression of the Arrhenius plot ln σ_{ac} versus $1000/T$ gives the slope of E_a values.

3.5 Magnetic Characterization of Nanocrystals

Electron motion on the atomic nucleus orbit is regulated by energy, angular momentum and spin. Energy is described by orbit radius around the nucleus. If it has larger energy, the orbit radius becomes larger. Angular momentum is a counterpart of integrated intensity. Spin corresponds to electron rotation. Magnetism of molecule depends on both orbital motion and spin [57]. The magnetic properties associated with the bulk materials are the same as that in magnetic properties of nanoparticles but due to the smaller size, magnetic nanoparticles show interesting properties. In magnetic properties of nanoparticles have some new phenomenon like Superparamagnetism and spin canting can be realized which may not be seen in bulk magnetic particles. The saturation magnetization of the nanomaterials is found to have less value and the coercivity of the nanoparticles is more when compared to bulk magnetic particles. The magnetic properties of as-prepared nanocrystals were investigated by using VSM at room temperature in field of up to 15 kOe.

3.5.1 Vibrating Sample Magnetometer (VSM)

The vibrating sample magnetometer or Foner magnetometer was developed by S. Foner in the 1950 [58]. VSM is a powerful technique to study the magnetic properties of materials easily, reliably and accurately, as the movement of the sample allows discrimination of the background signals. VSM is used to measure the magnetic properties (saturation, remanence, coercivity, anisotropy fields, etc) for a wide range of sample sizes and configurations, i.e., powders, solids, single crystals, thin films, and liquids, and also it is quite suitable for measurements at low, high, and ambient temperatures. Before the measurements, VSM has to be calibrated with a standard Ni to give a saturation magnetization of 55 ± 0.5 Am^2/kg at 0.5 T. For this, Ni standard is placed at the geometric center inside the pickup coils to perform a procedure known as "saddling" as shown in Fig. 3.15. In a standard set of electromagnets, the saddle point is located by finding the local extrema of the moment signal in the X, Y, and Z planes. In general, the magnetic moment is set to be a maximum in X and Z directions and minimum along the Y direction during saddling. The samples for analysis are placed with the same geometry like that of nickel standard [59].

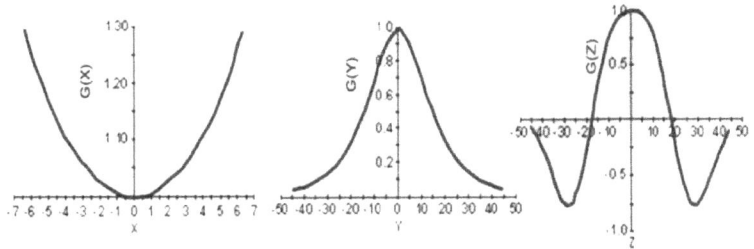

Fig. 3.15: Moment signals to find out the saddle point of the sample

The schematic of VSM is shown in Fig. 3.16. In the instrument, the sample to be measured is centered in the region between poles of the magnet, able to generate the measurable field H_0.

A thin vertical sample rod connects the sample holder with a transducer assembly located above the magnet. The transducers convert the sinusoidal ac driven signal into a sinusoidal vertical vibration of the sample rod. Thus, the sample is subjected to sinusoidal vibration in the uniform magnetic field H_0. In this case obtained readings vary only with the sample moment.

The Faraday's law states that an emf, V, is generated in a coil when there is a change in the magnetic flux linking the coil. For a coil with n turns of area, a, and magnetic flux density, B, the emf generated in the coil is:

$$V = -na\frac{dB}{dt} \qquad (3.49)$$

If the coil is positioned in a constant magnetic field, H, the magnetic flux density in the coil is given by,

$$B = \mu_0 H \qquad (3.50)$$

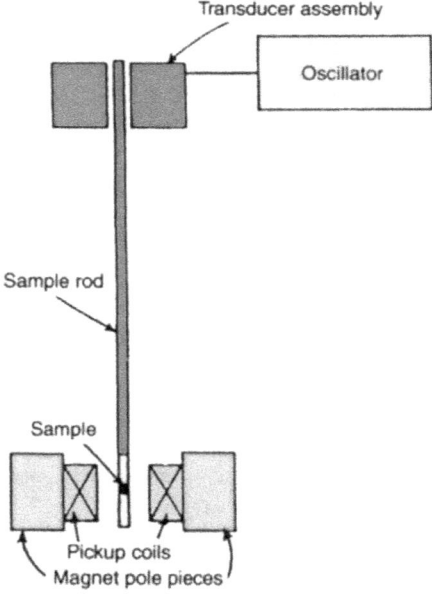

Fig. 3.16: Schematic of vibrating sample magnetometer

Consider a sample having magnetization, M, is introduced into the coil then, total magnetic flux in the system is,

$$B = \mu_0(H + M) \tag{3.51}$$

and the corresponding flux change is

$$\Delta B = \mu_0 M \tag{3.52}$$

Combining the equation (3.49) and (3.52) leads to,

$$Vdt = -na\mu_0 M \tag{3.53}$$

This means that the output signal of the coil is proportional to the magnetization M, but it is independent of the magnetic field in which the size of M is to be determined [60, 61].

Lake Shore Model: 7410 VSM of vibration frequency 82.5Hz and a maximum magnetic field of 15KGauss are used in the present study. The magnetic moment range of the instrument is 1μemu to 56emu.

3.5.2 Magnetic Properties

The effect of reducing the size of materials is of great importance from both fundamental considerations and modern practice. A brief discussion of magnetic behavior of low dimensional systems is focused based on literature. Magnetic nanoparticles exhibit specific properties such as coercivity and superparamagnetism, generally attributed to reduced dimensions.

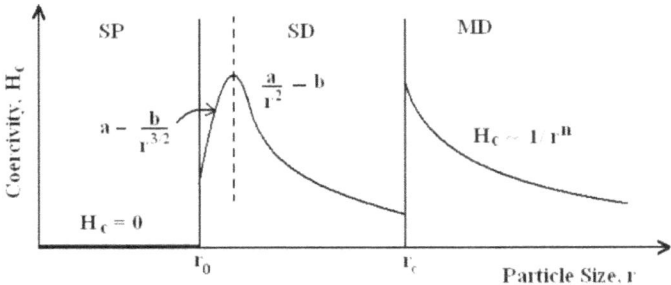

Fig. 3.17: Overview of the size dependence of coercivity exhibited by magnetic particles

3.5.2.1 Coercivity

The coercivity of fine particles has a striking dependence on their size. Fig. 3.17 shows very schematically, how the size range is divided, in relation to the variation of coercivity with particle radius r [62]. Beginning at larger size the following regions can be distinguished:

- Multi-domain (MD): It is observed for $r > r_c$ and in this region, the coercivity decreases as the particle size increases and the coercivity H_c is found to vary with size as $\sim 1/r^n$.
- Single-domain (SD): For $r_0 < r < r_c$, the particles become single domain and in this size range, the coercivity reaches a maximum.
- Superparamagnetic (SP): Below a critical size r_0, the coercivity is zero because of thermal effect, which is strong enough to spontaneously demagnetize the assembly of magnetic particles.

3.5.2.2 Paramagnetism

In certain materials, every molecule possesses a net stable magnetic moment due to orbital and spin magnetic moments still in the absence of an external magnetic field. It creates the net magnetic moment to zero, therefore the magnetization of the material is zero. The external magnetic field is applied to the magnetic dipoles tending to align themselves in the direction of the magnetic field with the material become magnetized. Although the paramagnetic material does not have the characteristics of the ferromagnetic material but they do have permanent dipole moments. Due to the thermal random motion the magnetic dipole moment does not line up perfectly with the applied magnetic field.

The susceptibility of paramagnetic material is of the order of 10^{-5} to 10^{-3} due to the absence of the strong magnetic effect as found in the ferromagnetic materials [63]. The paramagnetic material becomes a good magnetic material when it is placed in the strong magnetic material.

It starts acting as a magnet and attracts and repels other magnetic and ferromagnetic materials. As soon as the applied magnetic field is removed, the net magnetic alignment of the magnetic dipoles is lost and the dipoles return to their normal random motion. This consequence is recognized as Paramagnetism.

3.5.2.3 Superparamagnetism

Superparamagnetic substances consist of a small ferromagnetic cluster. A small ferromagnetic nanoparticle consists of a single domain. The internal forces and external magnetic field H determine the direction of its magnetization M. The magnetic moment carriers of this material are particles ($\mu \approx 10^4$ μB), which are composed of $\approx 10^4$ atoms, where μB is the Bohr magneton. There is no hysteresis for superparamagnetic materials [64]. The superparamagnetic materials always contain a particle size distribution. The behavior of a superparamagnetic material is often described by a modified Langevin function. The particle size distribution is required to explain the magnetic properties of the superparamagnetic system. For particles with spherical shape, the total magnetization can be evaluated by considering the Langevin function with a lognormal distribution [65]:

$$M / M_s = \int L(y) f(y) dy \qquad (3.54)$$

where L(y) is the Langevin function, M is the total magnetization of the sample, M_s is the saturation magnetization, f(y) is a log-normal distribution function with a median diameter value Dv and a standard deviation, where D is the diameter of the particle.

The magnetic properties of superparamagnetic materials are governed by the relative properties of the three energies:

- The anisotropic energy of particles.
- The Zeeman energy, which is due to the coupling between the magnetic field and the moment of the particles.
- The thermal energy, which tends to change the orientation of the magnetic moment in the individual particle.

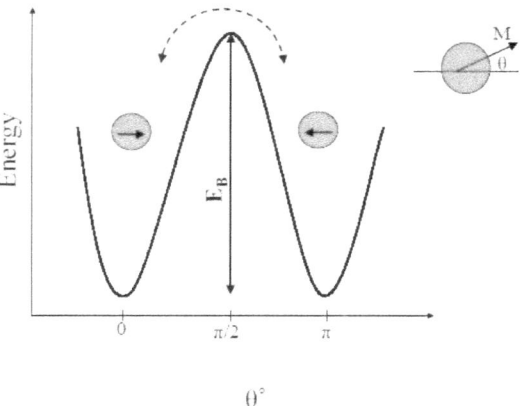

Fig. 3.18: Scheme of energy barrier separating states with opposite magnetization due to anisotropy

The ratios of these three energies determine which type of magnetic relaxation mechanism occurs. In the most common cases, where the magnetic anisotropy is unidirectional (even in spherical NPs defects favours a single easy magnetization axis), energy contains two minima states $(\theta = 0 \, and \, \theta = 180)$ with opposite magnetization (Fig. 3.18), separated by an energy barrier (ΔE). The magnetic relaxation can take place by two distinct mechanisms: The Neel relaxation in which, the direction of the moment of the particle is changed out of the easy direction against the magnetic anisotropy of the particle. The relaxation time τ_N for this process is given by Neel (Neel, 1949)

$$\tau_N = \tau_0 \exp(x) x^{-1/2} \, for \cdot x > 2 \qquad (3.55)$$

where $x = KV'/K_B T$. where V' is the magnetic volume of the particle, K_B is the anisotropy constant, $\tau_0 = 10^{-11} \, s.\tau_N$, and T is the temperature in Kelvin unit.

The other relaxation takes place through the rotation of the particle while the magnetic moment is held fixed by the magnetic anisotropy in the easy direction. This is known as Brownian relaxation and is given by (Neel, 1949):

$$\tau_B = 3V\eta_0 / k_B T \tag{3.56}$$

where η_0 is the viscosity of the carrier, and V is the hydrodynamic volume of the particle [66, 67].

3.5.3 Magnetic Anisotropy

The magnetic anisotropy is the internal energy of a system on the direction of the spontaneous magnetization. The energy of an atomic moment in the material depends on the orientation of this magnetic moment with respect the crystallographic axis. The exchange energy is established by the relative orientation of the moments with respect each other. Thus, the magnetic anisotropy energy contains contributions from several other sources. The dominant contribution comes from physical anisotropy such as crystal lattice, shape, stress and surface anisotropy that favors certain M orientations [68].

Magneto-crystalline anisotropy is due to the anisotropy of the crystal lattice, and the magnetic moment from the easy to the hard direction in a single crystal. The shape anisotropy is due to demagnetization effects in each direction of shape, strain anisotropy is due to magnetostriction. If the particle is subjected to stress, and surface anisotropy is due to the reduce symmetry of the surface sites. The resulting associated energy is, in the simplest case,

$$E_k(\theta) = KV \sin^2 \theta \tag{3.57}$$

where K is the effective anisotropy energy constant and is typically on the order of 10^3-10^5 J/m^3 in magnetic nanoparticles, V is the particle volume and θ is the angle between the magnetization and easy axis [69].

From this relation it is clear that the system has equal for $\theta = 0°$ or $180°$. However it requires energy KV to flip the magnetization vector from $\theta = 0°$ to $180°$. It is clear that the magneto-crystalline anisotropy energy is directly proportional to the volume of the crystal.

3.5.4 Effective Magnetic Moment

It is a convenient parameter (μ_{eff}) that is related to χ_m. The μ_{eff} is defined as:

$$(\mu_{eff})^2 = g^2 S(S+1)\beta^2 \quad (3.58)$$

If g=2, $\mu_{eff} = 2\sqrt{S(S+1)}$ \quad (3.59)

It can also be expressed in terms of number of unpaired electrons,

$$\mu_{eff} = 2\sqrt{S(S+1)} = \sqrt{n(n+1)} B.M \quad (3.60)$$

The μ_{eff} is related to the $\chi_M T$ values,

$$\mu_{eff} = \left(\frac{3k}{N\beta^2}\right)^{1/2} (\chi_M T)^{1/2} = \sqrt{8\chi_M T} = 2.83\sqrt{\chi_M T} \quad (3.61)$$

To recapitulate, one first makes a direct measurement of the volume susceptibility of a substance from which χ_M is calculated, and the above equation allows one to deduce the magnetic moment of the ion and molecule responsible for the paramagnetism [70].

3.6 Conclusions

MnS nanocrystals synthesized by wet chemical technique need to be characterized to assess the suitability of the crystal for various applications. In order to understand the behavior of nanostructures, the structural characterization is carried which investigates the crystalline perfection of the as grown sample.

This is followed by the optical characterization, which helps to check the transparency window, cut-off wavelength and optical band gap of the crystal. This result confirmed the materials for optoelectronic applications. In addition to the above studies, the dielectric, impedance, AC conductivity and magnetic behavior of the synthesized nanocrystals are also investigated. The instrumentation and characterization techniques used in the present study are discussed in this chapter.

References

1. Amanda J. Nichols, A Sol-gello Approach to the Synthesis of Nanocrystalline Materials, ProQuest, 2008.
2. Gourav Singla, K. Singh, Ceramics International 39 (2013) 1785-1792.
3. Xiangcheng Sun, A. Gutierrez, M. Jose Yacaman, Xinglong Dong, Shouri Jin, Materials Science and Engineering A286 (2000) 157-160.
4. Gus J. Palenik, William P. Jensen, Il-Hwan Suh, Journal of Chemical Education 80 (2003) 753-761.
5. A. W. Hull, Journal of the American Chemical Society 41 (1919) 1168-1175.
6. Vitaliz K. Pecharsky and Peter Y. Zavaliz, Springer, 2005
7. B. D. Cullity and S. R. Stock Elements of X-ray Diffraction, Prentice Hall, 2001.
8. W. I. F. David, K. Shankland, L. B. McCusker, Ch. Baerlocher; Structure Determination from Powder Diffraction Data, Oxford University Press, 2002.
9. E. Deydier, R. Guilet, S. Cren, V. Pereas, F. Mouchet, L. Gauthier, Journal of Hazardous Materials 146 (2007) 227–236.
10. D Lou, N Audebrand, Advances in X-ray Diffraction 41 (1997) 556-565.
11. Lawrence Kumar, Pawan Kumar, Amarendra Narayan, Manoranjan Kar, International Nano Letters 3 (2013) 1-8.
12. Vladimir Uvarov, Inna popov, Materials characterization 85 (2013) 111-123.
13. E Purushotham And N Gopi Krishna, Bulletin of Materials Science 37 (2014) 773–778.
14. Bharati R Rehani, P B Joshi, Kirit N Lad and Arun Pratap, Indian Journal of Pure & Applied Physics 44 (2006) 157-161.
15. VD Mote , Y Purushotham and BN Dole, Journal of Theoretical and Applied Physics, 6:6 (2012) 1-8.
16. R. J. Hill, Journal of Applied Crystallography 25 (1992) 589-610.
17. Gandhimathinathan Saroja, Veerapandy Vasu , Nagayasamy Nagarani, Journal of Metal 3 (2013) 57-63.
18. Yaroslav E. Romanyuk, Larysa P. Marushko, Lyudmyla V. Piskach, Ivan V. ityk, Anatolii O. Fedorchuk, Vasyl I. Pekhnyo, Oleg V. Parasyuk, CrystEngComm 15 (2013) 4838-4843.
19. I.E. Dubois, S. Holgersson, S. Allard, M.E. Malmström, Taylor & Francis Group, London, 2010.
20. H.Van Swygenhoven, Science 296 (2002) 66-67.
21. Y.P. Venkata Subbaiah, P.Prathap, K.T. Ramakrishna Reddy, Applied Surface Science 253 (2006) 2409-2413.
22. David B. Williams, C. Barry Carter, Transmission Electron Microscopy: A Textbook for Materials Science, 2009.
23. Anjam Khursheed, Scanning Electron Microscope Optics and Spectrometers, World scientific publishing company, 2011.
24. Jin Z. Zhang and Christian D. Grant, Annual review of nano research, world scientific publishing co, 2008.
25. David A. Puleo, Rena Bizios, Biological Interactions on Materials Surfaces, Springer science, 2009.
26. Kazuo Nakamoto, Infrared and Raman Spectra of Inorganic and Coordination Compounds, Wiley & Sons, England, 1984.
27. Dipankar Barpuzary and Mohammad Qureshi, Springer international publishing Switzerland, 2015.

28. Siladitya Behera, Subhajit Ghanty, Fahad Ahmad, Saayak Santra, and Sritoma Banerjee, Journal of Anal Bioanal Techniques 3 (2012) 1-6.
29. Fei Ke, Yu-Peng Yuan, Ling-Guang Qiu, Yu-Hua Shen, An-Jian Xie, Jun-Fa Zhu, Xing-You Tian and Li-De Zhang, Journal of Materials Chemistry 21 (2011) 3843-3848.
30. D. R. Vij, Luminescence of solids, Plenum Press, New York, 1998.
31. Ali Serpengüzel, Adnan Kurt, Ibrahim Inanç, James E. Cary, and Eric Mazur, Journal of Nanophotonics 2 (2008) 021770.
32. Gfroerer, T. H. Photoluminescence in Analysis of Surfaces and Interfaces, John Wiley & Sons, Ltd, 2006.
33. Fox. M, Optical Properties of Solids, Oxford University Press, 2004.
34. J. Ross Macdonald, Annals of Biomedical Engineering 20 (1992) 289-305.
35. Wei Lai and Sossina M. Haile, Journal of the American Ceramic Society 88 (2005) 2979–2997.
36. Dieter K. Schroder, Semiconductor Material and Device Characterization, John wiley & sons, 2006.
37. J. Ross Macdonald, Impedance spectroscopy, Wiley Newyork, 1987.
38. S Austin Suthanthiraraj and S Sarojini, Bulletin of Materials Science 36 (2013) 1297–1306.
39. Rajiv Ranjan, Nawnit Kumar, Banarji Behera, R. N. P.Choudhary, Advanced Materials Letters 5 (2014) 138-142.
40. Paul Hagenmuller, W. Van Gool, Solid Electrolytes: General Principles, Characterization, Materials, Applications, Elsevier, 2013.
41. B.A. Boukamp, Internal Report, CT 89/214/128.
42. Kao, K.C. Dielectric Phenomena in Solids, Elsevier Academic Press, London, 2004.
43. Chamarro, M. A., Gourdon, C., and Lavallard, P. Semiconductor Science and Technology 8 (1993) 1868–1874.
44. Pillai, O.S. Solid State Physics, New Age International, New Delhi, 2003.
45. Lixi, W.,Qiang, H., Lei, M., Qitu, Z. Journal of Rare Earths 25 (2007) 216-219.
46. Asghari Maqsood∗, Kishwar Khan, Journal of Alloys and Compounds 509 (2011) 3393–3397.
47. Navneet Singh, Ashish Agarwal , Sujata Sanghi, Satish Khasa, Journal of Magnetism and Magnetic Materials 324 (2012) 2506–2511.
48. M.D. Migahed, M. Ishra, T. Fahmy, A. Barakat, Journal of Physics and Chemistry of Solids 65 (2004) 1121-1125.
49. G. Williams, D.C. Watts, Transactions of the Faraday Society 66 (1970) 80-85.
50. P. B. Macedo, C. T. Moynihan, R. Bose, Physics and Chemistry of Glasses 13 (1972) 171-175.
51. W. C Hasz, C. T Moynihan, P. A Tick, J. Non-Cryst. Solids 172 (1994) 1363-1368.
52. E.K. Abdel-Khalek, Ibraheem Othman Ali, Journal of Non-Crystalline Solids 390 (2014) 31–36.
53. Suresh Sagadevan, Journal of Engineering Research and Applications 4 (2014) 126-130.
54. N Ponpandian, P Balaya and A Narayanasamy, Journal of Physics: Condensed Matter 14 (2002) 3221-3237.
55. Priyanka K P, Sunny Joseph, Smitha Thankachan, Mohammed E M, Thomas Varghese, Journal of Basic and Applied Physics 2 (2013) 4-7.

56. R. Selwin Joseyphus, E. Viswanathan, C. Justin Dhanaraj, J. Joseph, Journal of King Saud University - Science 24 (2012) 233–236.
57. Kiyoshi Nogi, Makio Naito, Toyokazu Yokoyama, Nanoparticle Technology Handbook, Elsevier, 2012.
58. István Meszaros, Journal of Electrical Engineering 63 (2012) 35-38.
59. Dodrill B.C, Low Moment Measurements with a Vibrating Sample Magnetometer, Technical report, LakeShore Cryotronics, 2010.
60. H. Kronmuller and S.S.P. Parken, Handbook of Magnetism and Advanced Magnetic Materials, Wiley, 2005.
61. K.H.J. Buschow, Concise Encyclopedia of Magnetic and Superconducting Materials, Elsevier, 2005.
62. R. C. O Handley, Modern magnetic materials principles and applications, John Wiley and Sons, 1999.
63. Nathan Ida, Engineering Electromagnetics, Springer, 2015.
64. D. G. Rancourt, & J. M. Daniels, Physical Review B 29 (1984) 2410.
65. T Kinoshita, S Seino, K Okitsu, T Nakayama, T Nakagawa, T.A Yamamoto, Journal of Alloys and Compounds 359 (2003) 46-50.
66. A. Sugawara, G.G. Hembree, M.R. Scheinfein: Journal of Applied Physics 82 (1997) 5662-5669.
67. J. Shen, R. Skomski, M. Klaua, H. Jenniches, S. Sundar Manoharan, J.Kirschner, Physical Review B 56 (1997) 2340.
68. Victor Kuncser, Lucica Miu, Size Effects in Nanostructures: Basics and Applications, Springer, 2014.
69. J. M. D. Coey, Magnetism and Magnetic Materials, Cambridge University Press, 2010.
70. Gregory S. Girolami, Thomas B. Rauchfuss, Robert J. Angelici, Synthesis and technique in inorganic chemistry, University Science Books, 1999.

Chapter IV

INFLUENCE OF REFLUXING TEMPERATURE ON THE STRUCTURAL, OPTICAL, ELECTRICAL AND MAGNETIC PROPERTIES OF CHEMICALLY SYNTHESIZED MnS NANOCRYSTALS

4.1 Introduction

Nanocrystalline metal sulfides as a typical and important group of semiconductors have attracted extensive investigation due to their unique optical, magnetic, and electrical properties. These nanoscale materials are expected to have many potential applications both in mesoscopic research electronics and development of nanodevices. Recently, a lot of effort has been focused on the magnetic semiconductors that exhibit both magnetic and useful semiconductor properties. The manganese chalcogenides (Sulfides, Selenides and Tellurides) are the magnetic materials and they have unique physical, morphological and chemical properties [1, 2]. These are paramagnetic compounds having five unpaired spins; owing to strong electron correlation and they do not form energy bands. During recent years, the dilute magnetic semiconductors (DMS) have been focused because they exhibit an interesting combination of magnetism and semiconductivity.

The manganese sulfide (MnS) semiconducting crystalline have attracted the attention of many researchers due to their potential applications in solar cell as a buffer material [3], solar selective coatings, opto-electronic devices [4, 5] like photo conductors and photo sensors, optical mass memories [6, 7, 8], antireflection coatings, spintronics applications and blue green light emitters [9]. The manganese sulfide is a wide band gap (Eg ≈ 3.8 eV) DMS material [10, 11] and it has outstanding magneto-optical properties [12]. Manganese sulfide (MnS) powder is in three different phases: α-MnS is the stable form (albandite) with rock-salt structure whereas β-MnS and γ-MnS are the metastable modifications with sphalerite and wurtzite structures, respectively [13]. Compared with the stable form, the β-MnS and γ-MnS are metastable forms which exhibit unique chemical properties as well as electrical, optical and magnetic properties [14, 15]. Various techniques such as Hydrothermal method [16, 17], Solvothermal synthesis [15, 18], Chemical bath deposition [4, 12], Successive ionic layer adsorption and reaction (SILAR) [19], Molecular beam epitaxy (MBE) [20], Radio-frequency sputtering [21, 22] and wet chemical synthesis (WCS) route [23] have been adopted for the fabrication of manganese sulfide (MnS). The wet chemical route exhibits several advantages like low cost, large-scale production, low-temperature process and no catalyst assistant.

This chapter presents the preparation of metastable MnS nanocrystals via wet chemical synthesis technique at low temperature. The influence of refluxing temperature during the preparation of MnS nanocrystals has been characterized. The structural, morphological, optical, electrical and magnetic properties have been investigated.

4.2 Experimental Details
4.2.1 Materials

Analytical grade Manganese acetate [$Mn(CH_3COO)_2$], Thioacetamide (CH_3CSNH_2), Ammonium chloride (NH_4Cl), Triethanolamine [$N(CH_2CH_2OH)_3$] and Trisodium citrate ($C_6H_5Na_3O_7$)

were purchased from Merck Chemicals, India and used as received without further purification. The analytical grade aqueous ammonia is used to adjust to the pH value of the solution. Distilled deionised water is used throughout the experiment.

4.2.2 Synthesis

The reaction scheme for the wet chemical synthesis of MnS nanocrystals is shown in Fig. 4.1. Aqueous solutions of manganese acetate [$Mn(CH_3COO)_2$] was used as manganese source and thioacetamide (CH_3CSNH_2) as sulphur source. The ammonium chloride as buffer, triethanolamine [$N(CH_2CH_2OH)_3$] and trisodium citrate ($C_6H_5Na_3O_7$) as complexing agents for preparing MnS nanocrystals. A typical synthesis was performed as follows: 1.0M manganese acetate was dissolved in 20ml of deionized water. Then 2ml of Triethanolamine followed by 20ml of 1.0M ammonium chloride were added drop by drop with continuous stirring. The solution became clear and homogenous. Afterwards 0.4ml of 0.7M trisodium citrate and 20ml of 1.0M thioacetamide solutions were introduced one by one into the mixed solution. The pH of solution was adjusted to 9.5 and stirred continuously for 2h. The solution was refluxed in a round bottom flask fitted in a heating mantle under vigorous stirring at 55°C, 65°C, and 75°C for 2h. Finally, the solution was cooled to room temperature and the synthesized nanocrystals were cautiously washed with ethanol, centrifuged, and finally dried at 60°C for 8h.

The precipitation of MnS nanocrystals occurs when the ionic product of Mn^{2+} and S^{2-} ions exceeds the solubility product of MnS. Thus the formation of MnS in this process can be written as follows:

$$Mn(CH_3COOO)_3 + TEA \rightarrow (Mn(TEA))^{2+} + 3CH_3COO^-$$

The overall reaction of formation of MnS can be written as:

$$Mn(TEA)^{2+} + CH_3CSNH_2 + 2OH^- \rightarrow MnS + TEA + CH_3CONH^2 + H_2O$$

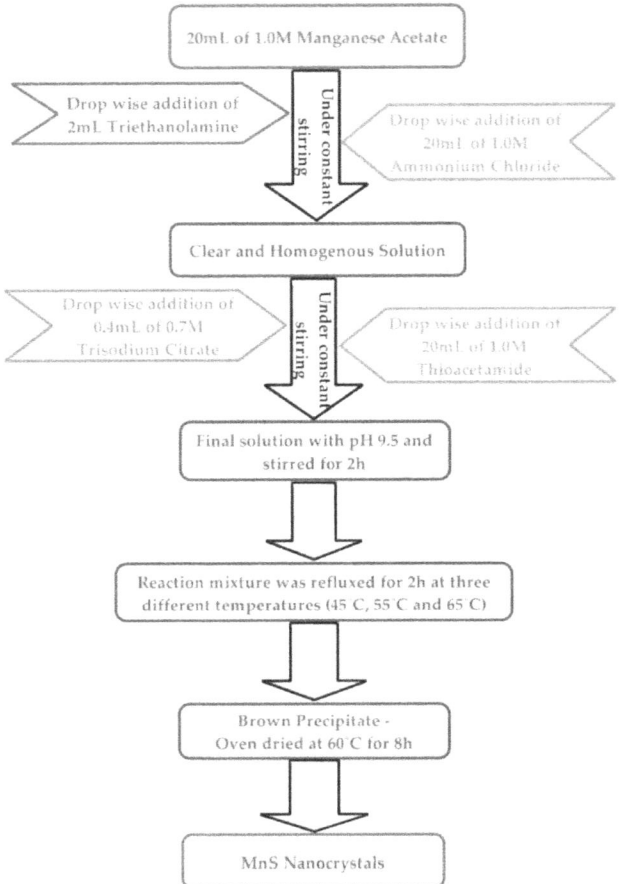

Fig. 4.1: Various steps involved in the wet chemical processing of MnS nanocrystals

4.3 Results and Discussion

4.3.1 Structural Properties

The crystalline structures of as-synthesized MnS nanocrystals have been investigated by X-ray diffraction and scanning electron microscopy.

4.3.1.1 XRD Analysis

In Fig. 4.2(a), the XRD pattern of MnS sample synthesized at 55°C refluxing temperature showed that the peaks at 2θ equal to 27.4°, 45.4° and 54.2° corresponding to the (111), (220), and (311) planes respectively, which match with diffraction of metastable sphalerite structured β-MnS with lattice constants a = b = c = 5.63A° (JCPDS 40-1288) [24]. The X-ray diffraction patterns of MnS samples shown in Fig. 4.2 (b, c) synthesized at 65°C and 75°C refluxing temperature exhibits three major peaks with (hkl) planes (100), (002) and (101) at around 2θ values of 25.7°, 27.7° and 45.6°, respectively. It can be well indexed to the wurtzite phase pure metastable γ-MnS (JCPDS: 40-1289) [25-27] with lattice constants a = 3.9A° and c = 6.41A°. It is shown that refluxing at more than 65°C brings to the phase transition from sphalerite to wurtzite.

The XRD patterns do not show significant peak shift but show noticeable broadening. The broad nature of the peaks indicates nanocrystalline nature of the synthesized samples. The degree of peak broadening decreases with increasing refluxing temperature which indicates the increase in crystallite size. From the Fig. 4.2, the peak intensity decreases with processing temperature increase, indicating decrease in crystallinity of the sample. This means that the MnS nanocrystals development is temperature dependent.

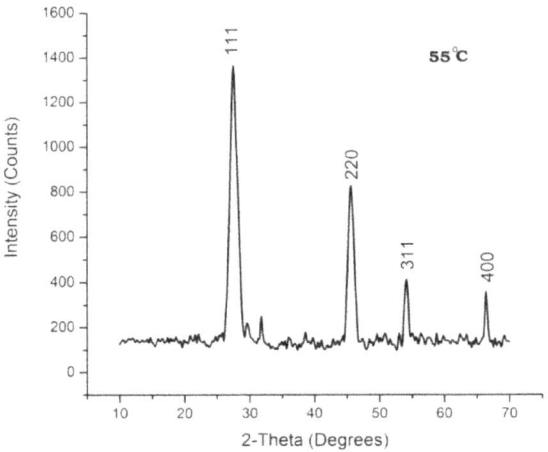

Fig. 4.2 (a): XRD pattern of MnS nanocrystals synthesized at refluxing temperature 55°C

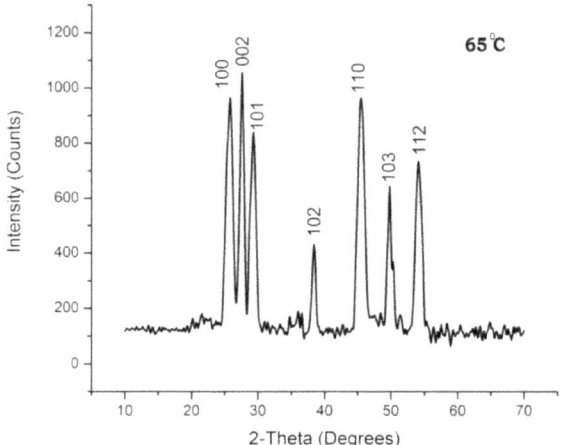

Fig. 4.2 (b): XRD pattern of MnS nanocrystals synthesized at refluxing temperature 65°C

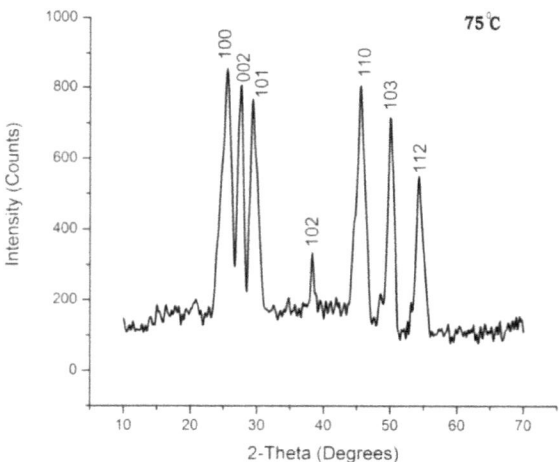

Fig. 4.2 (c): XRD pattern of MnS nanocrystals synthesized at refluxing temperature 75°C

The observed peak broadening β_0 can be represented as [28-29]

$$\beta_0 = \beta_i + \beta_r \tag{4.1}$$

where β_0 is the observed peak broadening in radius, β_i is the broadening due to instrumental factors in radians and β_r is the broadening due to crystallite size and lattice strain. The observed line breadth is simply the sum of Debye-Scherrer's formula and strain induced in powders.

$$\beta_r = \frac{K\lambda}{D\cos\theta} + 4\varepsilon\tan\theta \qquad (4.2)$$

By rearranging the equation 4.2, we get

$$\beta_r \cos\theta = \frac{K\lambda}{D} + 4\varepsilon\sin\theta \qquad (4.3)$$

The above equation is Williumson–Hall equation which is used to determine the crystallite size and the lattice strain of the as-prepared samples [30, 31]. In the equation, D is the crystallite size, K is a constant taken to be 0.94, β_r is the corrected full width at half maximum (FWHM) by subtracting the predetermined β_i for the instrument and λ is the wave length (0.15406 nm) of the X-rays.

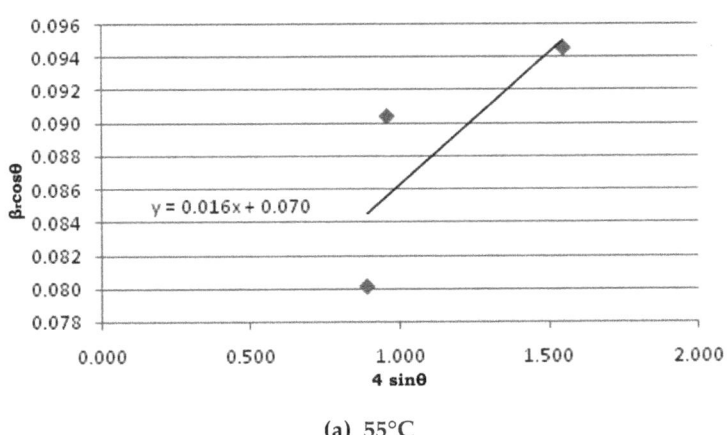

(a) 55°C

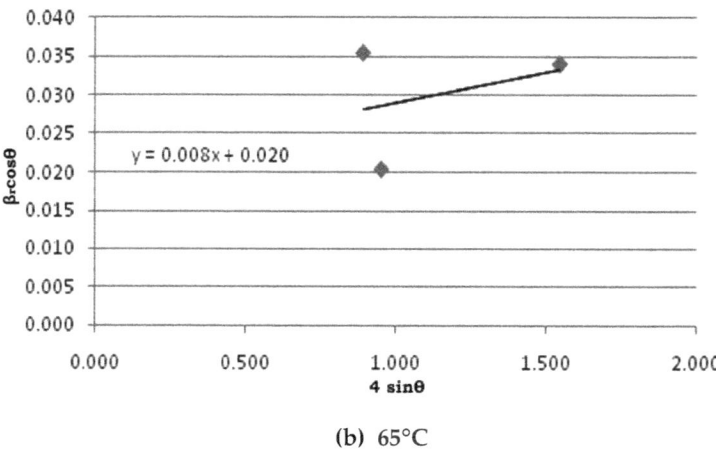

(b) 65°C

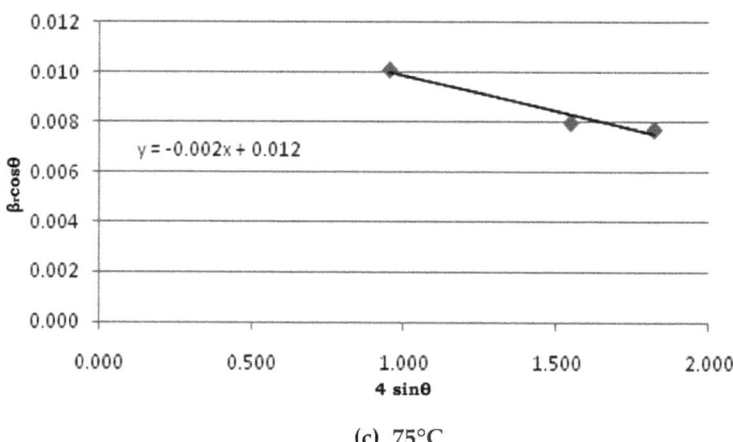

(c) 75°C

Fig. 4.3: W-H linear plots for the MnS nanocrystals synthesized at 55°C (a), 65°C (b) and 75°C (c) refluxing temperature to determine crystallite size and lattice strain

A plot is drawn with $4\sin\theta$ along the x-axis and $\beta_r \cos\theta$ along the y-axis for the MnS nanocrystals as shown in Fig. 4.3(a-c). From the linear fit to the data, the crystallite size is estimated from the y-intercept, and the stain (ε), from the slope of the fit. The variation of crystallite size and microstrain with refluxing temperature is shown in Table 4.1.

Table 4.1: XRD data analysis for Crystallite size, Micro strain and Dislocation density of MnS nanocrystals synthesized at different refluxing temperatures

Prepared Samples	2θ	β_o = FWHM in radians	$\beta r = \beta_o^2 - \beta_i^2$	$\ln(\beta_r)$	$\ln(1/\cos\theta)$	Crystallite size (D) in nm	Micro strain (ε)	Dislocation density (δ) (nm^{-2})
55°C	25.7000	0.082198	0.006756	-2.499	0.025	2.07	0.016	0.2336
	27.7000	0.093136	0.008674	-2.374	0.029			
	45.6000	0.102573	0.010521	-2.277	0.081			
65°C	25.8000	0.036494	0.001331	-3.311	0.026	7.24	0.008	0.0191
	27.6000	0.020933	0.000437	-3.867	0.029			
	45.4000	0.036982	0.001367	-3.298	0.081			
75°C	27.5000	0.010467	0.000109	-4.563	0.029	12.06	0.002	0.0069
	45.5000	0.008722	0.000075	-4.747	0.081			
	54.2000	0.008722	0.000075	-4.747	0.116			

The average crystallite sizes are calculated as 2.07 nm, 7.24 nm, and 12.06 nm for the sample synthesized at 55°C, 65°C and 75°C respectively. Micro strain and dislocation density are the crystal defect parameters [32]. The dislocation density of nanocrystals is given by $\delta = 1/D^2$. The micro strain (ε) and dislocation density (δ) of the as-prepared MnS samples are calculated. The microstrain is higher for the MnS prepared at low temperature, and increasing temperature the linear dependence of Williamson-Hall plot is weaker indicating that the broadening is due to crystallite size. Due to the increase in crystallite size with reflux temperature, the defects in the lattice are reduced, which in turn, reduces the microstrain. Also strain is an inherent and natural component of nanomaterials due to the large number of grain boundaries and the short spacing between them. Increase in particle size causes increase in surface energy which varies the magnitude of the strain. At higher reflux temperature, both the micro strain and the dislocation density are minimum, which reveals the reduction in the concentration of the lattice imperfections leading to preferred orientations.

The specific surface area with regard to the crystallite size is measured using the following relation [33],

$$D = \frac{6000}{S_{sp}\rho} \text{ m}^2\text{g}^{-1} \tag{4.4}$$

where S_{sp} is the specific surface area per unit mass of the sample and ρ is the density of MnS particles (3990000 gm^{-3}) [34]. The value of the specific surface area is determined as 729.9 m^{2}g^{-1}, 207.7 m^{2}g^{-1} and 124.6 m^{2}g^{-1} for the sample refluxed at 55°C, 65°C and 75°C, respectively. The specific surface area is often correlated with rates of dissolution and other phenomena such as catalyst activity, electrostatic properties, light scatting, opacity and many other properties which can influence the processing and behavior of powders. The results indicate decreases of the dislocation density with increasing the refluxing temperature, which leads to decreasing in specific surface area. This is due to the increase in crystallite size and small degree of agglomeration by the effect of temperature in the material.

4.3.1.2 SEM Analysis

The morphology of MnS powders originating from different refluxing temperatures (55°C, 65°C and 75°C) are revealed through SEM analysis. The SEM image in Fig. 4.4(a) shows that the MnS particles have nearly perfect spherical shapes with a smooth surface which is observed in the sample refluxed at 55°C. It would thus appear that the particle surface is particularly sensitive to the removal of the water that surrounding it. This sample has a great deal of spheres with different sizes which are assembled by small particles. The approximate diameter of as-synthesized MnS particles is 100 nm - 500 nm. The SEM image of MnS nanocrystals synthesized at the refluxing temperature of 65°C is shown in Fig. 4.4(b).

It exhibits spherical shape particles and size of the particles significantly increases as the refluxing temperature is increased. It is observed that textural mesopores are formed between the particles.

The diameter of the spherical MnS particles is about 500 nm - 1.5 µm. The SEM image also shows particles having a rough surface which may have an effect during the removal of water.

Fig. 4.4: SEM images of synthesized MnS nanocrystals refluxed at 55°C (a), 65°C (b) and 75°C (c)

The morphology of the product obtained by refluxing the synthesized solution at 75°C is shown in Fig. 4.4(c). It shows that the crystallinity of the sample decreased when compared to the sample synthesized at 55°C and 65°C refluxing temperature. The poor crystallinity of the MnS may lead to decrease in the carrier mobility. The particle size of the sample is about 2μm-3μm. The rise in refluxing temperature is not only alters the crystallite size but also demonstrate the surface texture belong to dendritic like morphology of the sample [35]. This observation reveals that the refluxing temperature is used to control the crystallization behavior of the MnS nanocrystals.

4.3.2 Optical Properties

The electronic state is an essential property that can be described in terms of valence, conduction bands and a gap between these bands. The wavelength of the electrons is nearer to the range of the particle sizes and the laws of classical physics have to be reserved by quantum confinement or quantum size effect. The optical properties of MnS nanocrystals are determined from absorbance measurements in the range of 200-800 nm. The PL studies illustrate different energy states available between valence and conduction bands responsible for radiative recombination. A material gains energy by absorbing photons and promotes electrons from a lower to higher energy level. It is described as making a transition from the ground state to an excited state of an atom, or from the valence band to the conduction band (CB) of a semiconductor. In a semiconductor, the energy difference between the filled valence band and the empty conduction band is of the order of a few electron volts. The optical band gap increases with a decreasing size in nanoscale range.

4.3.2.1 UV-Visible Absorption Spectra Analysis

The UV-Visible absorption spectra of MnS nanocrystals synthesized at different refluxing temperatures is shown in Fig. 4.5.

The linear nature of the plot shows that the mode of transition in this powder is of direct nature [36]. The peaks of the spectra correspond to the fundamental absorption edges in the samples, and could be used to estimate the optical band gap of the nanocrystals. The optical properties are strongly dependent on the particle size. It can be seen from the spectra that there is a uniform absorption in the visible range (800-390 nm). Absorption increases suddenly in the UV region. It is clear from the absorption spectra that an increase in absorption occurred at 287 nm, 295 nm and 301 nm for the sample synthesized at refluxing temperature 55°C, 65°C and 75°C respectively. The absorption edge is found at shorter wavelength for all the samples which is fairly blue shifted from the absorption edge of bulk MnS (326 nm) [37, 38].

The most direct way of extracting the optical band gap is to simply determine the photon energy at which there is sudden increase in the absorption. For bulk samples, the optical band gap is estimated from the $(\alpha h\nu)^2$ vs $h\nu$ plot, where α is the absorption coefficient and $h\nu$ is the photon energy. But for nanocrystalline sample the band gap is determined from absorption maxima. The optical band gap E_g (in eV) of the nanocrystalline sample is calculated from the absorption peak using the formula [39],

$$E_g = hc/\lambda \qquad (4.5)$$

where h is the Plank's constant, c is the velocity of light and λ is the wavelength at which absorption peak is obtained.

The calculated value of E_g for all the MnS nanocrystals are given in Table 4.2. It is observed that optical band gap value decreases from 4.32 eV to 4.12 eV as the refluxing temperature is increased from 55°C to 75°C. When grain size increases, the optical band gap energy is found to be decreased due to quantum confinement [40].

Table 4.2 Optical band gap of MnS nanocrystals synthesized at different refluxing temperatures

Sl. No.	Sample	Crstallite Size of as-synthesized MnS NC	Optical Band gap (Eg)
1	Bulk MnS	-	3.81 eV
2	MnS synthesized at 55°C refluxing temperature	2.07 nm	4.32 eV
3	MnS synthesized at 65°C refluxing temperature	7.24 nm	4.20 eV
4	MnS synthesized at 75°C refluxing temperature	12.06 nm	4.12 eV

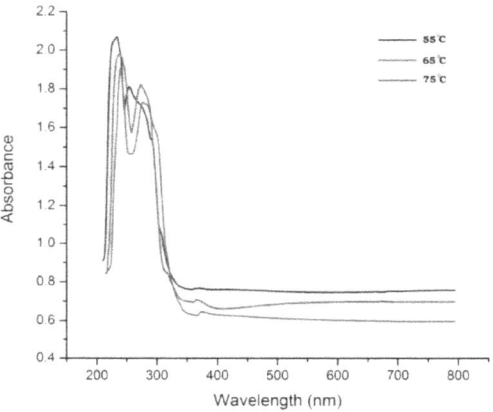

Fig. 4.5: Optical absorption spectra of MnS nanocrystals synthesized at different refluxing temperatures (55°C, 65°C and 75°C)

4.3.2.2 PL Spectra Analysis

Nanocrystals have discrete electron energy levels and surface defects play an important role in their electronic transitions [41]. Luminescence studies provide information regarding defects states, which take part in radiative de-excitation of the sample. Fig. 4.6 shows the room temperature photoluminescence (PL) spectra of as-prepared MnS nanocrystals with an excitation wavelength of 350 nm. It is clearly found that the PL spectra of each MnS sample exhibits two wide emission bands in blue region and the relative intensities of the two emission bands are different with the reaction temperature.

The emission bands may be attributed to the recombination of charge carriers in deep traps of surface localized states and a photo generated hole caused by surface defects [42, 43]. The mechanisms for blue emission from MnS are proposed with interstitial-manganese-related defect levels as initial states.

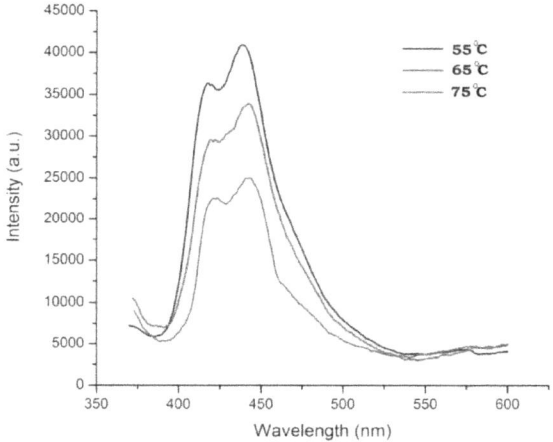

Fig. 4.6: Photoluminescence spectra of MnS nanocrystals synthesized at different refluxing temperatures (55°C, 65°C and 75°C)

The emission intensity decreases with the refluxing temperature from 55°C to 75°C. The higher intensity peaks centered at 410 nm and 438 nm for the refluxing temperature at 55°C. The sample synthesized at 65°C, the PL peaks centered at 413 nm and 440 nm have lower intensity compared to the sample synthesized at 55°C. When the refluxing temperature further increased to 75°C, the emissions observed at 416 nm and 443 nm shows less emission intensity compared to 55°C and 65°C. The lower PL intensity would indicate the less recombination of electron-hole pair. These results indicated that the peaks are slightly shifted to the red direction (higher wavelength) with the increase of refluxing temperature from 55°C to 75°C.

It is attributed to change in the relative position of the valance and conduction bands due to the dilatation of the lattice. Meanwhile, agglomeration can also lead to red shift in PL peak, according to the quantum size effect [44]. This agrees well with the results obtained from XRD and SEM observations. The strong emission lines of the sample at specific wavelength make it useful in optoelectronic applications [45].

4.3.3 Electrical Properties

Nanocrystalline materials exhibit enhanced electrical properties compared to their bulk crystalline counterparts. The dielectric behavior, electrical impedance, electric modulus, relaxation time, AC conductivity and activation energy of as-prepared MnS nanocrystals are analyzed. Studies on the effect of temperature and frequency on the dielectric, impedance and AC electrical conductivity offer valuable information about conduction phenomenon based on localized electric charge carriers in nanostructured materials.

4.3.3.1 Dielectric Studies

The dielectric constant (ε') of the samples are calculated from the measured values of capacitance using the formula [46]

$$\varepsilon' = \frac{C_p d}{\varepsilon_0 A} \tag{4.6}$$

where d is the thickness and A is the cross section area of the prepared pellet. ε_0 is the permittivity of free space. The variation of dielectric constant (ε') with frequency of the applied field for temperature ranges from 323K to 473K of samples synthesized at refluxing temperature of 55°C, 65°C, and 75°C are shown in Fig. 4.7(a, c & e). The dielectric constants, at all temperatures, are high at lower frequencies, which decrease as frequency increases and attain a constant low value at higher frequencies.

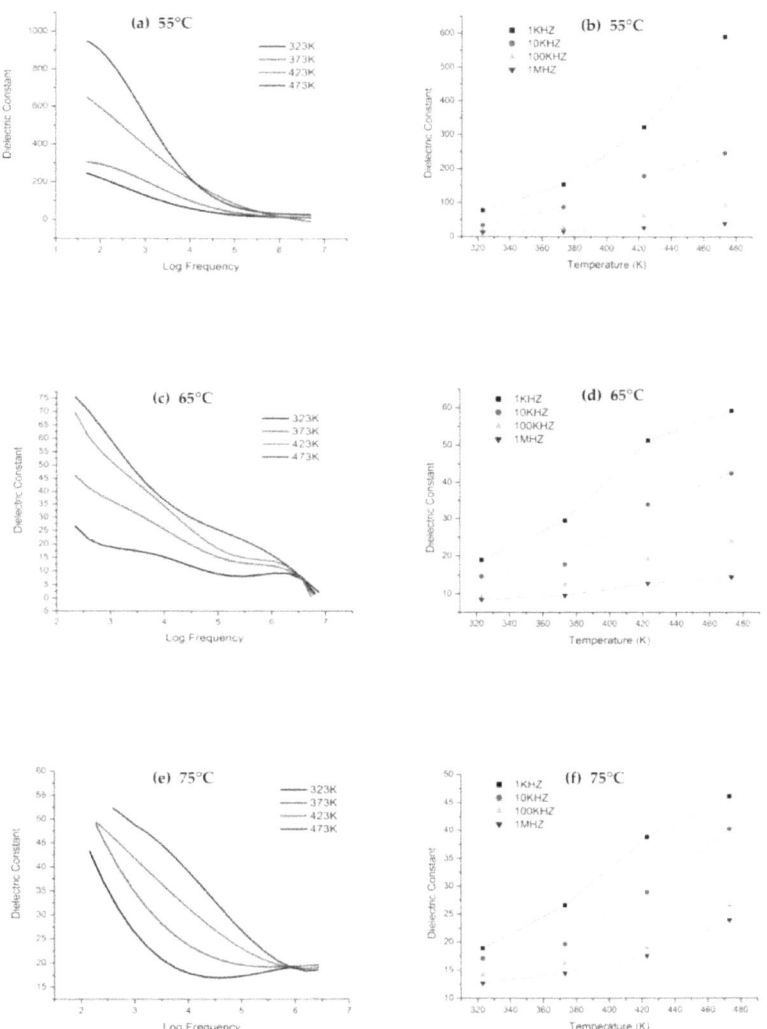

Fig. 4.7: Frequency and temperature dependence of dielectric constant for the MnS nanocrystals synthesized at 55°C (a & b), 65°C (c & d) and 75°C (e & f) refluxing temperature

In Fig.4.7 (a) for 323K, ε' is 980 at 100Hz for the sample refluxed at 55°C, which decreases to 70 at 1MHz. The corresponding variation at 473K is from 250 to 60. In Fig.4.7 (c) for 323K, ε' is 75 at 400Hz for the sample refluxed at 65°C, which decreases to 10 at 1MHz.

In the sample refluxed at 75°C (Fig. 4.7(e)), the corresponding variation is from 53 to 20 at 323K. The values of ε' decreases at low frequencies when the grain size is increased. The values of ε' at higher frequencies remain almost the same at an average of 10 to 20 for all the MnS samples. At high frequencies dielectric constant decreases owing to the reason of reduction of charge carriers. The dielectric constant (ε') decreases from 585 to 72 when the crystallite size increases from 2.06nm to 12.07nm at 323K and 1 KHz.

An electric field is applied, the space charges move and is trapped by defects resulting in the formation of dipole moments (space charge polarization). These dipoles will rotate, giving a resultant dipole moment in the direction of the applied field (rotation direction polarization) [47, 48]. Thus the high value of dielectric constant (ε') in the present study at low frequencies is mainly due to space charge polarization and rotation direction polarization. The dielectric constant initially decreases rapidly with increase in frequency which is attributed to the decrease in polarization. The dielectric constant remains fairly constant at higher frequencies for all samples due to the fact that beyond a certain frequency of the applied field, the electron exchange does not follow the alternating field. As temperature increases more and more dipoles will be oriented in the direction of the applied field, an increase in the value of dipole moment which increases dielectric constant with temperature. This is shown in temperature dependence graph Fig 4.7(b, d & f).

The variations of dielectric loss factor ($\tan \delta$) of MnS nanocrystals synthesized at different refluxing temperature with frequency are shown in Fig. 4.8 (a, c & e). It can be seen that $\tan \delta$ decreases with increase of frequency and at higher frequencies the loss angle has almost the same value at all temperatures. For 55°C sample, the value of the dielectric loss at 323K is 9.9 at 100Hz, which decreases to 0.2 at 1MHz. The corresponding variations of $\tan \delta$ for the samples synthesized at 65°C and 75°C are 9.6 to 0.4 and 9.5 to 2.1 respectively.

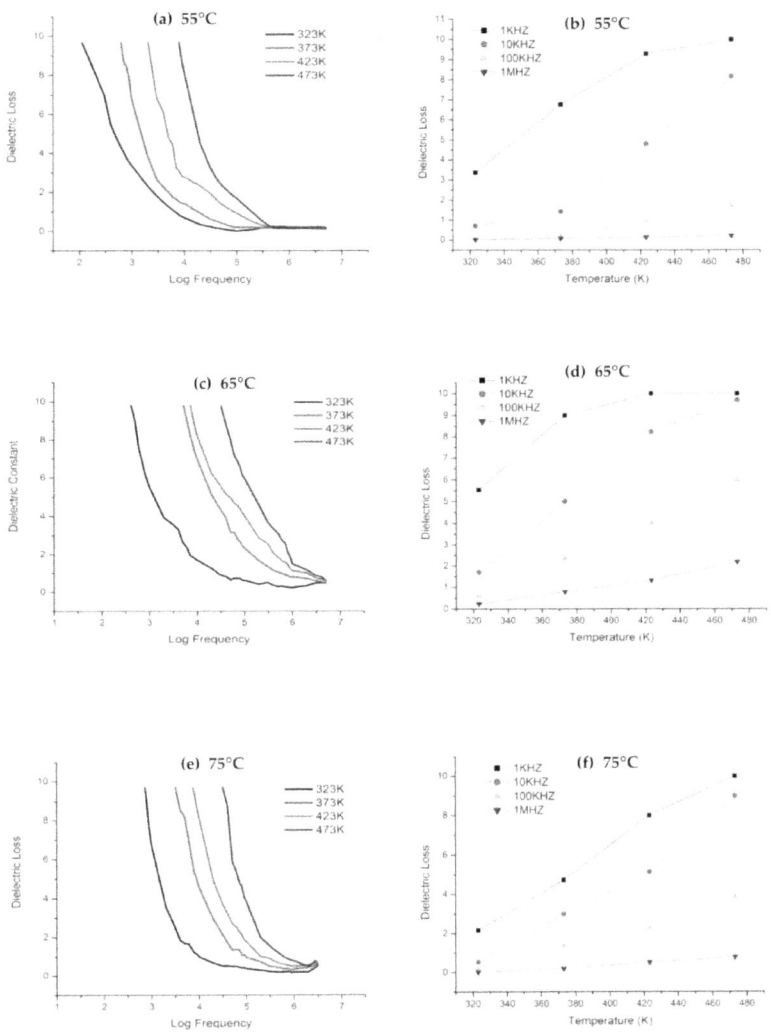

Fig. 4.8: Frequency and temperature dependence of dielectric loss for the MnS nanocrystals synthesized at 55°C (a & b), 65°C (c & d) and 75°C (e & f) refluxing temperature

The same behavior is also observed for other temperatures of 373K, 423K and 473K. It can be seen that the dielectric loss at 323K and at 100Hz decreases from 9.9 to 9.5 when the crystallite size increases from 2.06 nm to 12.07 nm. The variations of dielectric loss with temperature for fixed frequencies are also shown in Fig. 4.8(b, d & f). In Fig 4.8(b), dielectric loss at 1 KHz increases from 3.4 to 9.9. The corresponding variations are observed in Fig 4.8 (d) & (f). For all the samples, $\tan \delta$ initially increases with temperature, attains a high value and then slightly decreases. It is also observed that dielectric loss increase more rapidly at lower frequencies than higher frequencies. At low frequencies, the rapid increase of $\tan \delta$ due to the interfacial and dipolar polarization [49].

4.3.3.2 Impedance Studies

Electrical impedance describes a measure of opposition to alternating current. Electrical impedance extends the concept of resistance to AC circuits, describing not only the relative amplitudes of the voltage and current, but also the relative phases. When the circuit is driven with direct current (DC), there is no distinction between impedance and resistance. In impedance spectroscopy, the impedance of a sample is obtained as a function of the frequency measuring the response of the sample to an applied excitation signal. Normally, a small AC voltage is applied over a range of frequencies and measured the electrical current response. The complex impedance is then calculated as [50]

$$Z = \frac{V}{I} \qquad (4.7)$$

where V is the applied voltage and I is the measured current. As the impedance is a complex quantity, it can be expressed separating the real and imaginary parts,

$$Z = Z' - iZ'' \qquad (4.8)$$

or calculation of the magnitude and the phase angle:

$$|Z| = \sqrt{(Z')^2 + (Z'')^2}$$

(4.9)

$$\angle Z = -\arctan\left(\frac{Z''}{Z'}\right)$$

(4.10)

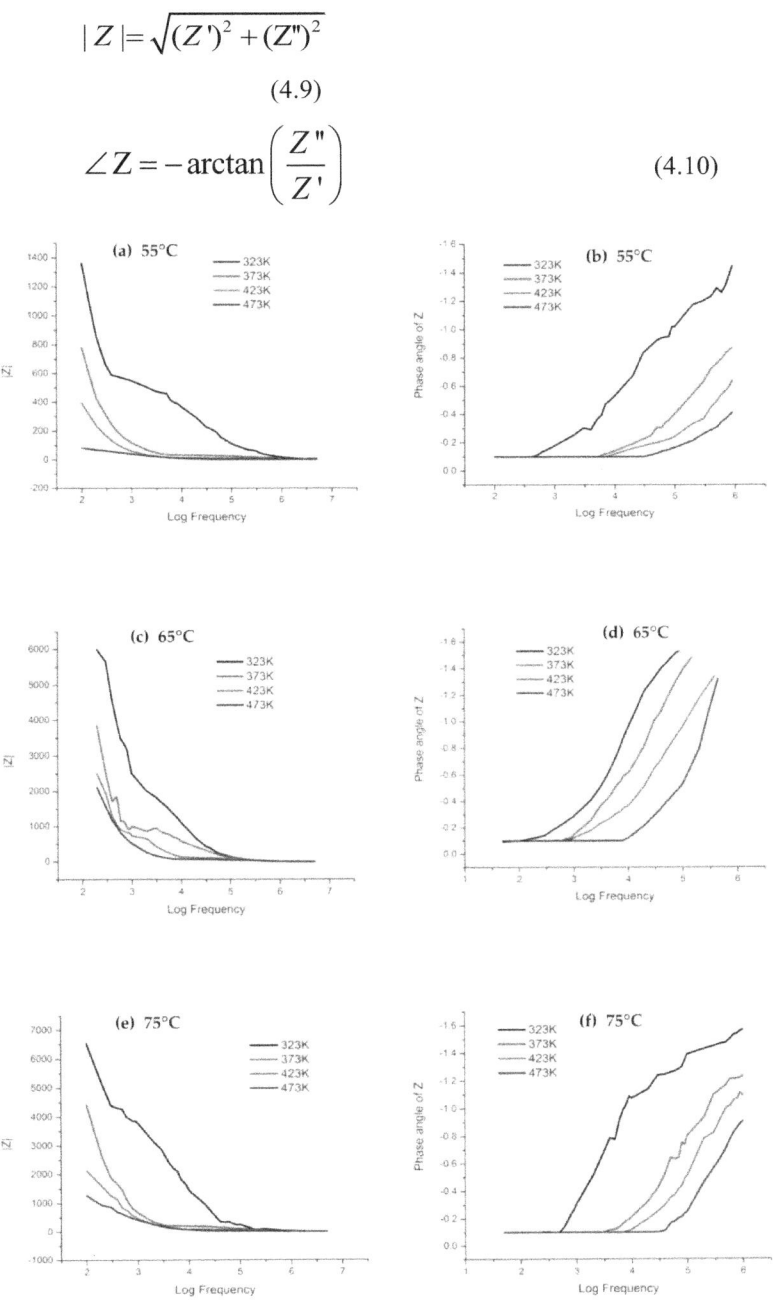

Fig. 4.9: Variation of the impedance magnitude and phase angle as function of frequency for the MnS nanocrystals synthesized at 55°C (a & b), 65°C (c & d) and 75°C (e & f) refluxing temperature

The electrical impedance spectrum (modulus |Z| and phase θ with frequency) of each sample has been studied. Fig. 4.9(a-f) shows the impedance magnitude and phase spectra of MnS nanocrystals synthesized at three different refluxing temperatures (55°C, 65°C and 75°C). The plots show that the magnitude of the impedance is affected by the variation in refluxing temperature in the frequency range 100 Hz to 5 MHz. The maximum magnitude variations are observed in between 100 Hz and 10 KHz. The impedance magnitude |Z| is increased when increase the refluxing temperature from 55°C to 75°C which is caused by decreasing conductivity of sample. At low frequency the complex impedance values are higher which indicate the larger polarization, which on further increase in frequency shows independent behavior.

The impedance magnitude values for all temperatures merge at high frequency. This behavior is due to the release of space charge as a result of reduction in barrier properties [51]. Also, it may be a responsible factor for the enhancement of AC conductivity with temperature at high frequencies. Moreover, at low frequency |Z| values decreases with increase in temperature exhibiting negative temperature coefficient of resistance (NTCR) type behavior similar to that of semiconductors [52]. The phase angle decreases with increase in frequency for all the MnS samples are shown in Fig 4.9(b, d & f).

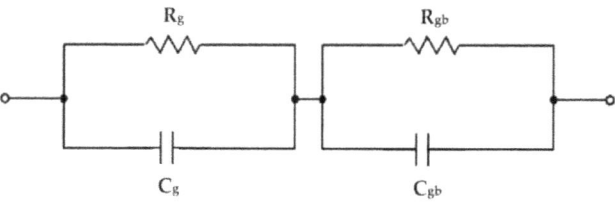

Fig. 4.10: Equivalent circuit model of as-prepared MnS nanocrystals from impedance analysis

The Cole-Cole plot or Nyquist is drawn between Z" data and Z' data. The experimental impedance spectra (Cole-Cole plot) are modeled by an ideal equivalent circuit consisting of a resistor R and a capacitor C.

Since crystalline materials generally show grain and grain-boundary impedances, they can be represented by the equivalent circuit [53]. The circuit shown in Fig. 4.10 consists of a series array of two sub-circuits that represents grain and grain boundaries effects. Each sub-circuit is composed of a resistor and capacitor connected in parallel. Let R_g, R_{gb} and C_g, C_{gb} be the resistances and capacitances of grains and grain boundaries respectively.

The Cole-Cole plot of MnS nanocrystals synthesized at 55°C, 65°C and 75°C refluxing temperature are shown in Fig. 4.11(a-c). The semicircular arcs are clearly observed at all studied temperatures in which the small semicircle occurring (not visible) in the high frequency region corresponds to grain resistance while the large semicircle in the low frequency region corresponds to grain boundaries. This indicates the contribution of both grain and grain boundary in conduction mechanism [53].

The prepared samples are taken for impedance analysis at different temperature range from 323K to 473K. The grain and grain boundary resistances decreases as the intercept of semicircular arc on real axis reduced with the rise in temperature. It suggests that with increase in temperature the bulk conductivity increases, which is a typical behavior of semiconductors [54].

In Fig 4.11 (a), R_{gb} reduces from 40 KΩ to 3 KΩ as temperature rises from 323K to 473K. In Fig 4.11 (b), the corresponding variation is from 86 KΩ to 14 KΩ. In Fig 4.11 (c), the corresponding variation is from 310 KΩ to 27 KΩ. The same behavior is also obtained in other temperatures (373K, 423K and 473K). The grain and grain boundary parameters like resistance and capacitance are obtained by analyzing the impedance data which are shown in Table 4.3. The value of R_{gb} at 323K increases from 40 KΩ to 310 KΩ when the grain size of the sample increases from 2.07 nm to 12.06 nm.

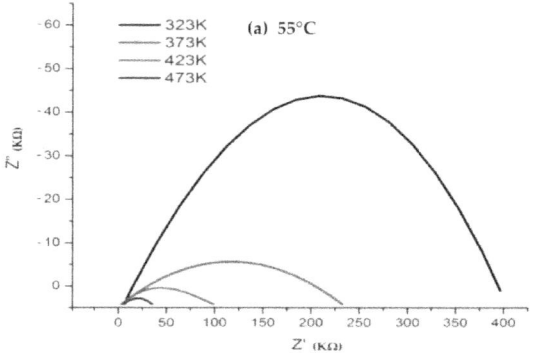
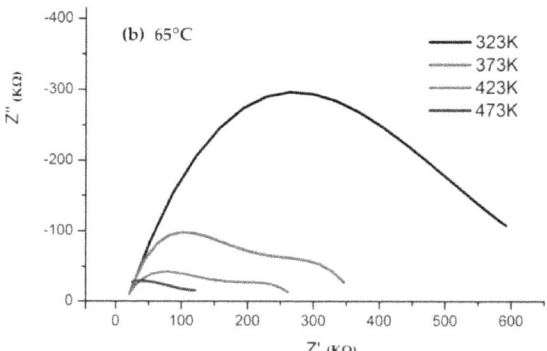
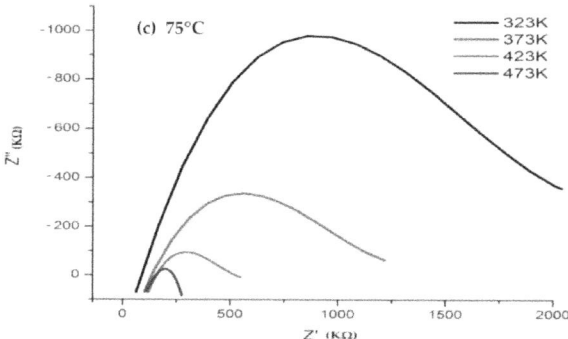

Fig. 4.11: Cole-Cole plots for the MnS nanocrystals synthesized at 55°C (a), 65°C (b) and 75°C (c) refluxing temperature

Table 4.3: Equivalent circuit parameters of impedance spectra for the MnS nanocrystals synthesized at different refluxing temperatures

Samples and Factors		Grain resistance (R_g) Ω	Grain boundary resistance (R_{gb}) KΩ	Grain capacitance (C_g) F	Grain boundary capacitance (C_{gb}) F
55°C Refluxing Temperature	323K	322	40	9.17E-08	7.38E-07
	373K	271	23	1.47E-08	1.73E-07
	423K	247	9	7.20E-09	1.98E-07
	473K	202	3	1.97E-09	1.33E-07
65°C Refluxing Temperature	323K	407	86	1.23E-07	5.83E-07
	373K	386	38	8.19E-08	8.32E-07
	423K	371	26	3.39E-08	4.84E-07
	473K	354	14	1.03E-08	2.59E-07
75°C Refluxing Temperature	323K	1337	310	1.19E-07	5.11E-07
	373K	1242	147	3.21E-08	2.71E-07
	423K	904	66	2.78E-08	3.81E-07
	473K	589	27	1.35E-08	2.94E-07

4.3.3.3 Electric Modulus Studies

The electrical response can be analyzed through complex electric modulus formalism. The complex electric modulus (M*) is calculated by using the relation

$$M^*(\omega) = M'(\omega) + jM''(\omega) = j\omega C_0 Z^*(\omega) \quad (4.11)$$

where,

$$M' = \omega C_0 Z''$$
$$M'' = \omega C_0 Z'$$
$$C_0 = \varepsilon_0 A / t$$

where ε_0 is permittivity in free space, A is area of electrode surface and t is thickness of the sample [55]. Fig. 4.12(a-f) shows the angular frequency dependence of M'(ω) and M''(ω) for the MnS nanocrystals as a function of temperature. It also shows that the modulus peaks shift towards higher frequency side on increasing temperature. In the low temperature region, the value of M'(ω) increases with the increase in frequency, while in the high-temperature region the value of M'(ω) increases rapidly with the increase in both the temperature and frequency.

It may be contributing to the conduction phenomena due to the short range mobility of charge carriers. This implies the lack of a restoring force for flow of charge under the influence of a steady electric field [56].

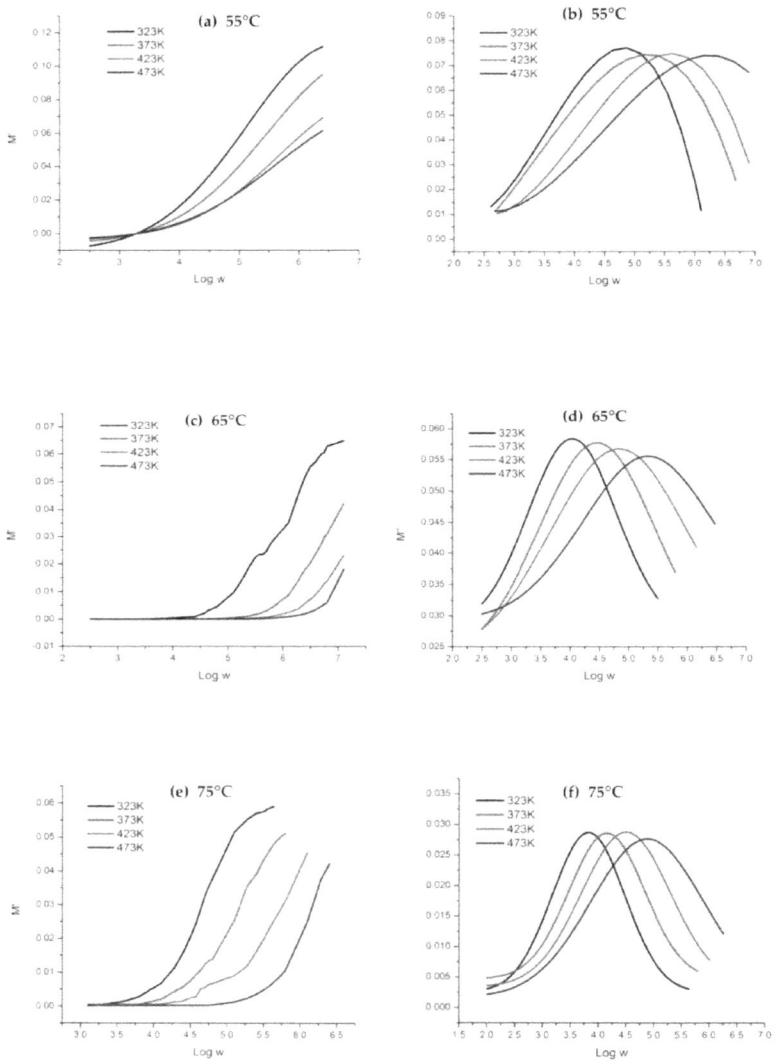

Fig. 4.12: Variation of M' and M'' with frequency at different temperatures for the MnS nanocrystals synthesized at 55°C (a & b), 65°C (c & d) and 75°C (e & f) refluxing temperature

It may be noted that the variation of M" with frequency attains a maximum value at a particular frequency, and that peak is shifted to higher frequency with rise in temperature. The peak shift is attributed to long range motion of ions that are thermally activated. The low frequency side of the peak represents the range of frequency in which the ions can move long range, i.e., ions can perform hopping from one site to the neighboring site. The high frequency side of the M'' represents the range of frequencies in which the ions are separately confined to their potential wells, and the ions can make only the localized motion within the well [57]. The peak in M" vs frequency plot is asymmetric in nature representing the spread of relaxation time. The asymmetric nature of the modulus peak indicates the stretched exponential character of relaxation time and hence the relaxations of non-Debye type [57]. The frequency ω_m (M" max) gives the most probable relaxation time τ_m from the condition $\omega_m \tau_m = 1$. This temperature dependent of relaxation time follows the Arrhenius behavior which is shown in Fig. 4.13.

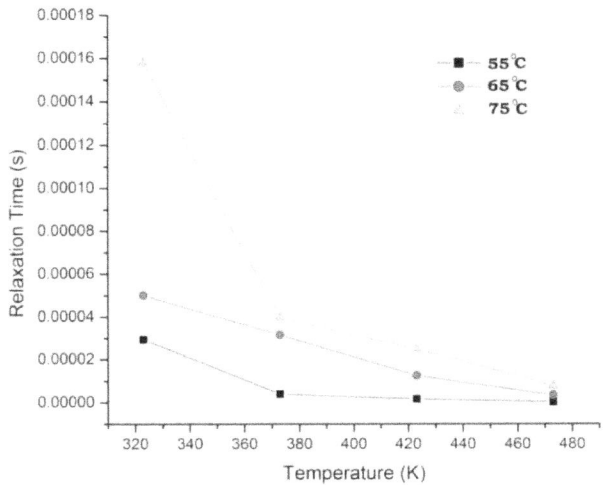

Fig. 4.13: Relaxation time (τ) of as-prepared MnS nanocrystals synthesized at 55°C, 65°C and 75°C refluxing temperature

With increasing the refluxing temperature, the M" peak position is shifted towards lower frequency side; it indicates that the relaxation time increases in the material. Also, M" peak frequencies increase with increase in temperature and the shift in frequency of M" peaks correspond to the conductivity relaxation.

4.3.3.4 AC Conductivity Studies

The AC conductivity peaks with frequency at different temperatures are shown in Fig. 4.14 (a-c). The frequency dependence conductivity of the material exhibits both low and high frequency dispersion phenomena. The AC conductivity is obtained from the dielectric constant (ε') and loss tangent ($\tan\delta$) using the relation [58]

$$\sigma_{ac} = \varepsilon'\varepsilon_0 \omega \tan\delta \qquad (4.12)$$

where ω is the angular frequency and ε_0 is the permittivity of vacuum. Thus AC conductivity depends strongly on the frequency of the applied field. At 323K temperature, σ_{ac} has a low value, which increases steadily and reaches the maximum value at 473 K for all samples. It is evident that the AC conductivity increases with the increase in temperature and frequency, indicating the mobility of charge carriers responsible for hopping [59].

The increasing conductivity with temperature is associated with the enhancement in the drift mobility and hopping frequency of charge carriers with increasing temperature according to the relation

$$\sigma = ne\mu \qquad (4.13)$$

where n is the total number of charge carriers, e is the electronic charge and μ is the mobility of charge carriers [59]. Fig. 4.15(a-c) shows the variation of σ_{ac} against the $10^3/T$ of as-prepared MnS nanocrystals at different frequencies. The conductivity versus temperature response can be explained by a thermally activated transport of Arrhenius type,

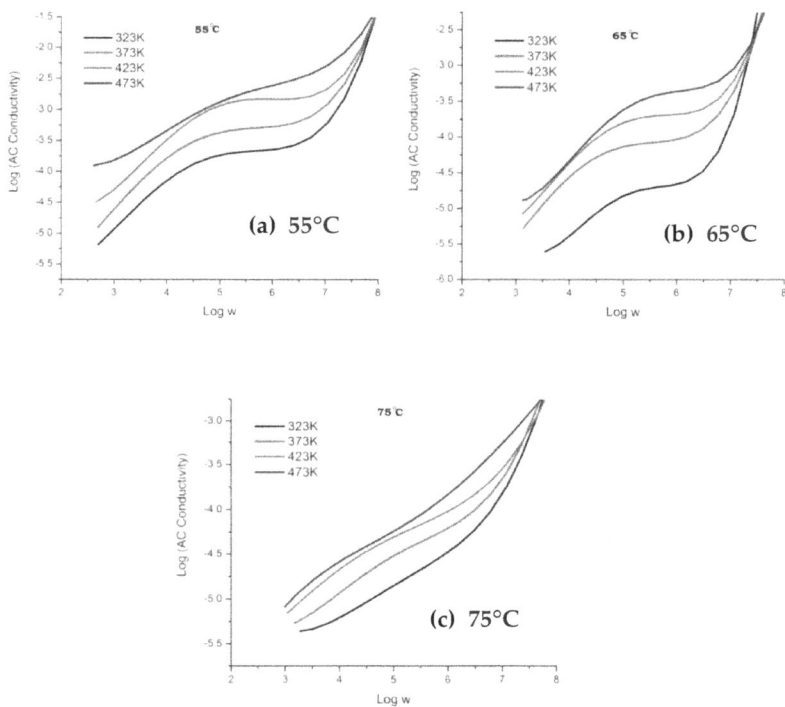

Fig. 4.14: Frequency dependence of the AC conductivity (σ_{ac}) for the MnS nanocrystals synthesized at 55°C (a), 65°C (b) and 75°C (c) refluxing temperature

$$\sigma_{ac} = \sigma_0 \exp\left(\frac{-E_a}{k_B T}\right) \qquad (4.14)$$

where σ_0 is the pre exponential term, k_B is the Boltzmann's constant, and E_a is the activation energy of the mobile charge carriers [60].

At lower temperature, a small deviation from the linear behavior of conductivity is noticed and can be attributed to Mott's hopping type phenomena. The electric conductivity is controlled by the migration of charge species under the action of electric field and by the defect-ion complexes, the polarization field, the relaxations, etc.

At lower temperatures, the defect ion complexes may exist and they are present in a polarization field. The polarization field may aid the charge species to migrate, thereby contributing to the conductivity. However, the defect ion complexes may cause distortions [61]. The grain boundaries may also contribute to the conduction mechanism. The defect ion complexes tend to dissociate. The activation energy may be due to the migration energy for the charges.

In Fig. 4.16 at 3KHz, there are distinct values of activation energy E_a are found to be 0.27 eV, 0.34 eV and 0.38 eV for the samples synthesized at 55°C, 65°C and 75°C refluxing temperature respectively. The conductivity-based activation energies show a decreasing trend of variation with increasing frequency.

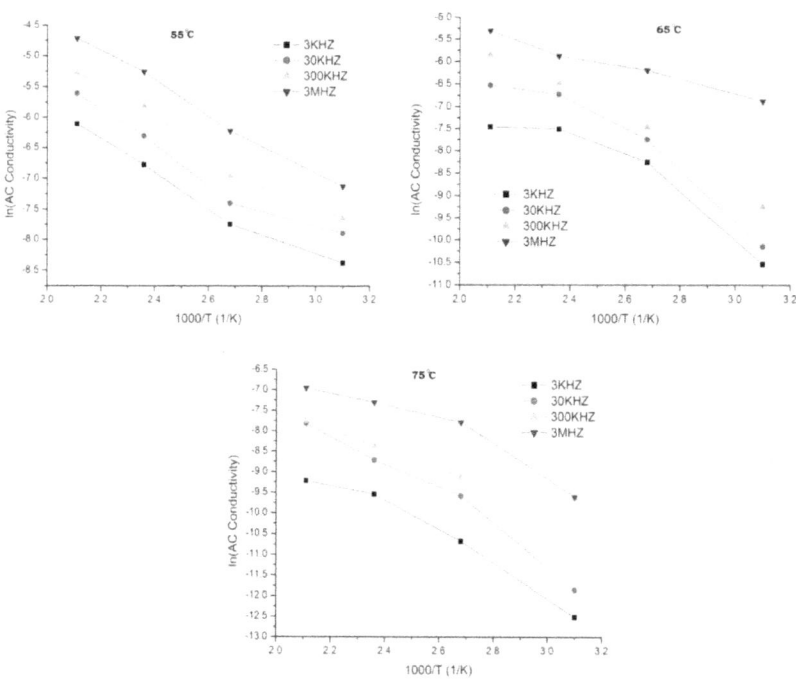

Fig. 4.15: Variation of σ_{ac} against the $10^3/T$ of the MnS nanocrystals synthesized at 55°C (a), 65°C (b) and 75°C (c) refluxing temperature

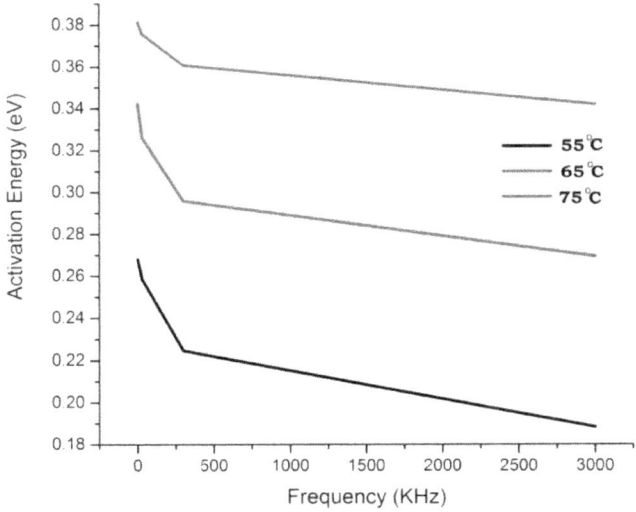

Fig. 4.16: Variation of activation energy with frequency for the MnS nanocrystals synthesized at 55°C, 65°C and 75°C refluxing temperature

4.3.4 Magnetic Properties

Room temperature magnetic measurements of entire samples are carried out in the field range of 15kOe which is shown in Fig. 4.17. An M-H plot of refluxing temperature at 55°C, 65°C and 75°C samples do not show hysteresis, however they exhibits a linear nature. The saturation magnetization (M_s) and the Coercivity (H_c) of the as-prepared powders are found to be 0.788 emu/g and 4.39 G for 55°C, 0.945 emu/g and 4.16 G for 65°C, and 0.985 emu/g and 4.09 G for 75°C, respectively. With increasing particle size the amount of specific surface area decreases, leading to increase in saturation magnetization because of the decreasing non magnetic surface layer. The coercivity values of MnS particles decreased with increase in refluxing temperature due to magnetization reversal from domain wall motion. The coercivity and remanent magnetization are negligible values, which indicate that all the samples exhibit paramagnetic behavior.

The corresponding squareness ratios (M_r/M_s) are found to be 0.000318, 0.000304 and 0.000292 for the sample synthesized at 55°C, 65°C and 75°C refluxing temperature, respectively. The value of M_r/M_s is <<0.01 suggesting the mixture of super paramagnetic grains (SP) with some multi domain grains (MD) [62, 63].

The susceptibility is calculated from the slope of a linear fit of the M vs. H experimental data. The susceptibility and permeability are 5.26E-5 emu/g.Oe & 1.25607E-06, 6.31E-5 emu/g.Oe & 1.25608E-06, and 6.55E-5 emu/g.Oe & 1.25609E-06 for the sample synthesized at 55°C, 65°C and, 75°C refluxing temperature respectively. The both permeability and magnetic susceptibility are increases with an increasing concentration of magnetic grains. Paramagnetism results the presence of atoms with permanent magnetic dipoles or atoms with unpaired electrons. A magnet formed by the spinning of the negatively charged particle on its axis due to single electron.

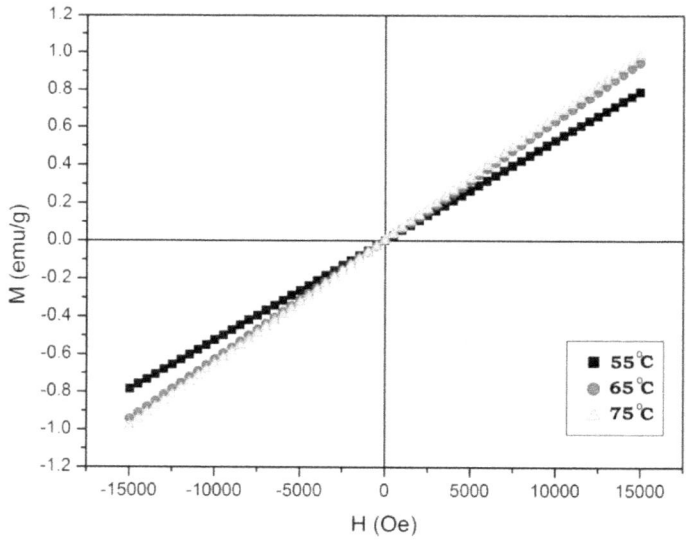

Fig. 4.17: Magnetic moment vs. magnetic field plots of the MnS nanocrystals synthesized at 55°C, 65°C and 75°C refluxing temperature

An electron traveling in a closed path around the nucleus produces a magnetic moment. The magnetic moment μ_s [64] of a single electron is given from wave mechanics as:

$$\mu_s(in\mu_B) = g\sqrt{s(s+1)} \qquad (4.15)$$

where s is the spin quantum number and g is the gyromagnetic ratio [65]. The spin magnetic moment of one electron is calculated to be 5.91 μ_B from the electron spin alone. The effective magnetic moment of ions inside the MnS are also estimated using

$$\mu_{eff} = \sqrt{3k_B T \chi / N} \qquad (4.16)$$

where χ is magnetic susceptibility at temperature T, $k_B = 1.38 \times 10^{-16}$ erg K^{-1}, and N is Avogadro number. The obtained effective magnetic moment μ_{eff} is 1.48 μ_B, 1.55 μ_B, and 1.58 μ_B which is much less than that of full magnetic moment of 5.91 μ_B [66] with L=0 and S=5/2. One can estimate K_{eff} and particle volume by using the following relations [44-45].

$$K_{eff} = \frac{H_c M_s}{2} \qquad (4.17)$$

$$V = \frac{25 K_B T_B}{K_A} \qquad (4.18)$$

where, K_{eff} and K_A are effective magnetic anisotropy constant [67], H_C is coercivity, M_S is saturation magnetization, K_B is Boltzmann constant (1.38×10^{-23} J/K), T_B is blocking temperature in K, V is particle volume. The estimated values are,

K_{eff} = 172.78 J.m^{-3} and V = 9.08E-21 m^3 for 55°C refluxing temperature
K_{eff} = 220.33 J.m^{-3} and V = 7.11E-21 m^3 for 65°C refluxing temperature
K_{eff} = 82.21 J.m^{-3} and V = 1.90E-20 m^3 for 75°C refluxing temperature.

The exchange interaction between particles may take place, which tends to align the magnetic moments of neighboring particles parallel and overcome the magneto-crystalline anisotropy and the demagnetization effect of individual particles.

The energy of magneto static interaction (dipole-dipole) depends on the magnetic moment is given by [68].

$$U_{max} = \frac{-\pi^2 D^3 M_S^2}{18} \qquad (4.19)$$

There is equal magnetic moment when the dipoles are parallel. The calculated energy of magneto static interactions are -7.37E-14, -1.35E-13, and -3.08E-13 for the sample refluxed at 55°C, 65°C, and 75°C respectively. According to Neel, the magnetic moment of the particle relaxes to its equilibrium position with a relaxation time τ_N [69] given by the relation,

$$\tau_N = \tau_0 \exp\left(\frac{K_{eff} V}{k_B T}\right) \qquad (4.20)$$

The Neel relaxation in which, the direction of the moment of the particle is changed out of the easy direction against the magnetic anisotropy of the particle. The relaxation time τ_N for this process is $\sim 1 \times 10^{-9}$ s for all as-synthesized MnS nanocrystals.

4.4 Conclusions

The manganese sulfide nanocrystals synthesized at three different refluxing temperature (55°C, 65°C and 75°C) conditions using wet chemical synthesis route and their structural, optical, electrical and magnetic characterization have been studied. The sphalerite to wurtzite phase transformation of MnS nanocrystals is observed in the reaction solutions refluxed at 65°C. X-ray peak broadening analysis is used to evaluate the crystalline sizes and lattice strain by the Williamson-Hall analysis. It is observed that the strain value decreased but the crystallite size increased as refluxing temperature of the samples are increased. The average crystallite size of MnS nanocrystals is found to be 2.07 nm, 7.24 nm and 12.06 nm for the sample refluxed at 55°C, 65°C and 75°C respectively.

The specific surface area value decreased from 729.9m^2/g to 124.6m^2/g with an increase in the refluxing temperature from 55°C to 75°C. The SEM micrograph revealed that the spherical MnS particles in the range of 100nm–3μm. The dendritic like structure is observed in the response to increasing refluxing temperature to 75°C.

UV-Visible absorption spectra showed a strong absorption peak at around 287 nm (4.32 eV) for 55°C, 297 nm (4.20 eV) for 65°C and 301 nm (4.12 eV) for 75°C refluxing temperature, which are considerably blue-shifted compared to that of bulk phase MnS (3.81 eV). The PL studies showed strong and stable blue emissions and decrease in emission intensities with increase in refluxing temperature. In electrical studies, the dielectric properties and AC conductivity of MnS nanocrystals are investigated as a function of frequency and temperature.

The dielectric studies reveal that both the dielectric constant and dielectric loss decrease with an increase in the frequency and it remains constant at higher frequencies. The grain and grain boundary contributions are studied using Cole-Cole impedance plot and it is found that both the grain and grain boundary resistance decrease with rise in operating temperature. The long tail of M' in the low frequency range indicates the capacitive nature of the system. The M'' spectra peaks showing the non-Debye type conductivity relaxation. The frequency dependent AC conductivity at different temperatures indicates that the conduction process is thermally activated process. Long range mobility of charge carrier are playing effective role in conduction process at elevated temperature. The activation energy increases from 0.27 eV to 0.38 eV on increasing the refluxing temperature from 55°C to 75°C. VSM measurements of as-synthesized MnS nanocrystals evidenced all the samples presented a strong paramagnetic behavior which can be attributed to the presence of magnetic dipoles located on the surface of nanocrystals that exhibited a minimum interaction with their neighbors inside of the crystals.

References

1. S.H. Wei, A. Zunger, Phys. Rev. B 48 (1993) 6111.
2. G.R. Wu, K. Nagatomo, M. Sasaki, F. Nagasakki, H. Sato, M. Taniguchi, W.X. Gao, Solid State Commun. 118 (2001) 425-429.
3. C.D. Lokhande, A. Ennaoui, P.S. Patil, M. Giersig, M. Muller, K. Diesner, H. Tribursch, Thin Solid Films 330 (1998) 70-75.
4. S. Mochizuki, B. Piriou, J. D. Ghys, Journal of Physics: Condensed Matter 2 (1990) 5225-5229.
5. T. Dietl, H. Ohno, F. Matsukura, J. Cibert, D. Ferrand, Science 287 (2000) 1019-1022.
6. Tappero. R, D'Arco. P, Lichanot. A, Chemical Physics Letters 273 (1997) 83-90.
7. D Hobbs, J Hafner, J. Phys.: Condens. Matter 11 (1999) 8197.
8. A.N. Kravtsova, I.E. Stekhin, A.V. Soldatov, X. Liu, M.E. Fleet, Physical Review B 69 (2004) 134109.
9. R.L Gunshor, A.V. Nurmikko (Eds.), Semiconductor and Semimetals 44 (1997) 1-58.
10. Ning Zhang, Ran Yi, ZhongWang, Rongrong Shi, HaidongWang, Guanzhou Qiu, Xiaohe Liu, Materials Chemistry and Physics 111 (2008) 13-16.
11. O. Goede, W. Heimbrodt, Physica status solidi (b) 146 (1988) 11-62.
12. R.R. Galazka, J. Cryst. Growth 72 (1985) 364-370.
13. YongCai Zhang, Hao Wang, Bo Wang, HaiYan Xu, Hui Yan, Masahiro Yoshimura, Optical Materials 23 (2003) 433-437.
14. Jin Mu, Zhenfang Gu, Lei Wang, Zhiqing Zhang, Hua Sun, Shi-Zhao Kang, Journal of Nanoparticle Research 10 (2008) 197-201.
15. Jun Lu, Pengfei Qi, Yiya Peng, Zhaoyu Meng, Zhiping Yang, Weichao Yu, Yitai Qian, Chemistry of materials 13 (2001) 2169-2172.
16. Pingtang Zhao, Qiumei Zeng, Xianliang He, Hao Tang, Kaixun Huang, Journal of Crystal Growth 310 (2008) 4268-4272.
17. Changhua An, Kaibin Tang, Xianming Liu, Fanqing Li, Guien Zhou, Yitai Qian, Journal of Crystal Growth 252 (2003) 575-580.
18. Xinhua Zhang, Yiqing Chen, Chong Jia, Qingtao Zhou, Yong Su, Bo Peng, Song Yin, Minjun Xin, Materials Letters 62 (2008) 125-127.
19. H. M. Pathan, S. S. Kale, C.D. Lokhande, S.H. Han, O.S. Joo, Materials Research Bulletin 42 (2007) 1565-1569.
20. M. Okajima, T. Tohda, Journal of Crystal Growth 117 (1992) 810-815.
21. Sandra A. Mayen-Hernandez, Sergio Jimenez-Sandoval, Rebeca Castanedo-Perez, Gerardo Torres-Delgado, Benjamin S. Chao, Omar Jimenez-Sandoval, Journal of Crystal Growth 256 (2003) 12–19.
22. I.Oidor-Juarez, Jimenez, G Torres-Delgado, R Castanedo-Perez, O Jimenez-Sandoval, B Chao, S Jimenez-Sandoval, Mater. Res. Bull. 37 (2002) 1749-1754.
23. L.P.S. Santos, E.R. Camargo, M.T. Fabbro, E. Longo, E.R. Leite, Ceramics International 33 (2007) 1205-1209.
24. F. M. Michel, M. A. A. Schoonen, X. V. Zhang, S. T. Martin, J. B. Parise, Chem. Mater., 18 (2006) 1726-1736.
25. Subhajit Biswas, Soumitra Kar, Subhadra Chaudhuri, Journal of Crystal Growth, 299 (2007) 94-102.
26. Meiying Liu, Nannan Shan, Linlin Chen, Xiaoqian Li, Bona Li, Wansheng you, Applied surface science 258 (2012) 7922-7927.

27. YongCai Zhang, Hao Wang, Bo Wang, HaiYan Xu, Hui Yan, Masahiro Yoshimura, Optical Materials 23 (2003) 433-437.
28. B.R. Rehani, P.B. Joshi, K.N. Lad, A. Pratap, Indian J. Pure Appl. Phys. 44 (2006) 157-161.
29. Hiten Sarma, K.C. Sarma, International Journal of Scientific and Research Publications 4 (2014) 1-7.
30. VD Mote, Y Purushotham, BN Dole, Journal of Theoretical and Applied Physics 6:6, (2012) 1-8.
31. Khorsand Zak, Abd. Majid, W.H, Abrishami, M.E., Yousefi, R, Solid State Sciences 13 (2011) 251-256.
32. K. Venkateswarlu, M. Sandhyarani, T.A. Nellaippan, N. Rameshbabu, Procedia Materials Science 5 (2014) 212-221.
33. A. Monshi, M.R. Foroughi, M.R. Monshi, World J. Nanosci. Eng. 2 (2012) 154-160.
34. V.A.Rodrigues, W.A.Monteiro, A.M.S-Silva, N.A.M.Ferreira, L.C.E.Silva, Congresso Brasileiro de Engenharia e Ciencia dos Materiais, 2006.
35. Jian Yin, Xue Han, Yanping Cao, Conghua Lu, Sci Rep. 4 (2014) 5710-5723.
36. Nada K.Abbas, Khalid T. Al- Rasoul , Zainb J. Shanan, Int. J. Electrochem. Sci. 8 (2013) 3049-3056.
37. Xin Yu, Cao Li-yun, Huang Jian-feng, Liu Jia, Fei Jie, Yao Chun-yan, Journal of Alloys and Compounds 549 (2013) 1-5.
38. Subhajit Biswas, Soumitra Kar, Subhadra Chaudhuri, Jour. of Crys. Growth 284 (2005) 129-135.
39. Preeti Gupta, M. Ramrakhiani, The Open Nanoscience Journal 3 (2009) 15-19.
40. H. Lin, C.P. Huang, W. Li, C. Ni, S. Ismat Shah, Yao-Hsuan Tseng, Applied Catalysis B: Environmental 68 (2006) 1-11.
41. Andrew M. Smith, Shuming Nie, Acc Chem Res. 43 (2010) 190-200.
42. J.P. Borah, K.C. Sarma, Acta Phys. Polonica A 114 (2008) 713-719.
43. M. Bhagwat, P. Shah, V. Ramaswamy, Mater. Lett. 57 (2003) 1604-1611.
44. Zhijun Wang, Feng Tao, Feng Pan, Yufeng Sun, Weili Cai, Lianzeng Yao, Applied Surface Science 258 (2011) 44-49.
45. Mou Pal, N.R. Mathews, Erik R. Morales, J.M. Gracia y Jiménez, X. Mathew, Optical Materials 35 (2013) 2664-2669.
46. M. S. Gaur, Ajay Pal Indolia, Purushottam Kumar, Advances in Polymer Technology 32 (2012) 1-13.
47. S. Suresh, C. Arunseshan, Appl Nanosci 4 (2014) 179-184.
48. Nisha J Tharayil, R Raveendran, Alexander Varghese Vaidyan, PG Chithra, Indian journal of Engineering & Materials Sciences 15 (2008) 489-496.
49. Yang, Peter Kofinas, Polymer 48 (2007) 791-798.
50. M.K. Sharma, R.N. Gayen, A.K. Pal, D. Kanjilal, Ratnamala Chatterjee, Solid State Communications 151 (2011) 1182-1187.
51. Subhanarayan Sahoo, Umasankar Dash, S. K. S. Parashar, S. M. Ali, Journal of Advanced Ceramic 2 (2013) 291-300.
52. Piyush R. Das, B. Pati, B. C. Sutar, R. N. P. Choudhury, Journal of Modern Physics 3 (2012) 870-880.
53. Khalid Mujasam Batoo, Shalendra Kumar, Chan Gyan Lee, Alimuddin, Journal of Alloys and Compounds 480 (2009) 596-602.
54. Seok-Hyun Yoon, Clive A. Randall, Kang-Heon Hur, Journal of the American Ceramic Society 92 (2009) 1758-1765.

55. Taha A. Hanafy, Advances in Materials Physics and Chemistry 2 (2012) 255-266.
56. Z. Anwar, M. Azhar Khan, I. Ali, M. Asghar, M. Sher, I. Shakir, M. Sarfraz, M. Farooq Warsi, Journal of Ovonic Research 10 (2014) 265-273.
57. Ghada E. El-Falaky, Osiris W. Guirguis, Nadia S. Abd El-Aal, Progress in Natural Science: Materials International 22 (2012) 86-93.
58. D.R. Patil, S.A. Lokare, R.S. Devan, S.S. Chougule, C.M. Kanamadi, Y.D. Kolekar, B.K. Chougule, Materials Chemistry and Physics 104 (2007) 254-257.
59. Khalid Mujasam Batoo, Shalendra Kumar, Chan Gyu Lee, Alimuddin, Current Applied Physics 9 (2009) 1072-1078.
60. Bernard A. Boukamp, Mai T.N. Pham, Dave H.A. Blank, Henny J.M. Bouwmeester, Solid State Ionics 170 (2004) 239-254.
61. Md. T. Rahman, M. Vargas, C.V. Ramana, Journal of Alloys and Compounds 617 (2014) 547-562.
62. T. Kurz, L. Chen, F. J. Brieler, P. J. Klar, H.A. Krug von Nidda, M. Froba, W. Heimbrodt, A. Loidl, Phys. Rev. B 78 (2008) 132408.
63. Thirugnanasambandan, Theivasanthi, Marimuthu Alagar, Physics and technical sciences 1 (2013) 39-45.
64. Carolyn I. Pearce, Richard A.D. Pattrick, David J. Vaughan, Reviews in Mineralogy and Geochemistry 61 (2006) 127-180.
65. Felicia Iacomi, M. Vasilescu, S. Simon, Surface Science 600 (2006) 4323-4327.
66. Remy Tappero, Albert Lichanot, Chemical Physics 236 (1998) 97-105.
67. S. Shafiu, R. Topkaya, A. Baykal, M.S. Toprak, Materials Research Bulletin 48 (2013) 4066-4071.
68. Soshin Chikazumi, Physics of magnetism, New York, Wiley, 1964.
69. Yu. P.Kalmykov, W. T. Coffey, S. V. Titov, Physics of the Solid State 47 (2005) 272-280.

Chapter V

INFLUENCE OF Mn/S MOLAR RATIO ON THE STRUCTURAL, OPTICAL, ELECTRICAL AND MAGNETIC PROPERTIES OF CHEMICALLY SYNTHESIZED MnS NANOCRYSTALS

5.1 Introduction

Among metal sulfide nanocrystals, MnS is gaining popularity, because it exhibits completely new and enhanced properties. At nanoscale range, the particle size leads to large surface area per mass where a large number of atoms are in immediate contact and available for reaction. The size, morphology and properties (chemical, physical and electrical) of the metal nanocrystals are strongly influenced by the experimental conditions [1, 2]. Hence, the variations of the experimental conditions including capping molecules, precursor molar concentration, and reaction temperature/aging have been explored to control the size and shape of the nanocrystals. A variety of methods have been reported for the preparation of MnS nanocrystals, but the wet chemical synthesis route is simple, low cost method and has a large scale production potential [3]. It can be successfully used to obtain uniform MnS nanocrystalline powders at low temperature. Aqueous solutions of manganese acetate and thioacetamide were used as manganese source and sulfur source respectively.

The stoichiometric molar ratio of thioacetamide to manganese acetate is favorable to produce the metastable sphalerite and metastable wurtzite MnS phase with strongly reduced dimensions allows for the possibility of realizing controlled quantum confinement as a step toward quantum engineering. The triethanolamine and trisodium citrate were used as complexing agents for the preparation of MnS nanocrystals. The purpose of complexing agents is to ensure the slow release of Mn^{2+} ions during the MnS preparation. It is well known from ancient times that MnS nanocrystals are very useful for anode material for Li-ion batteries [4], biomedicine and optoelectronic devices [5], buffer material in solar cell [6] and magneto-optical devices [7].

This chapter presents the influence of Mn/S molar ratio in precursor on the structural, electrical, optical and magnetic properties of MnS nanocrystals synthesized via wet chemical route. X-ray diffraction, Scanning microscopy, Fourier transform infrared spectroscopy, UV-Visible spectroscopy, Photoluminescence spectroscopy, Impedance spectroscopy and Vibrating sample magnetometer were carried out to study these properties of manganese sulfide nanocrystals.

5.2 Experimental Details

5.2.1 Materials

For the preparation of MnS nanocrystals, the materials used are manganese acetate [$Mn(CH_3COO)_2$], thioacetamide (CH_3CSNH_2), ammonium chloride (NH_4Cl), triethanolamine [$N(CH_2CH_2OH)_3$] and trisodium citrate ($C_6H_5Na_3O_7$). All chemicals used are of analytical grade purchased from Merck Chemicals, India and used as received without further purification. Deionized water is used for all the preparation process.

5.2.2 Synthesis

A schematic diagram of the formation of MnS nanocrystals is shown in Fig. 5.1.

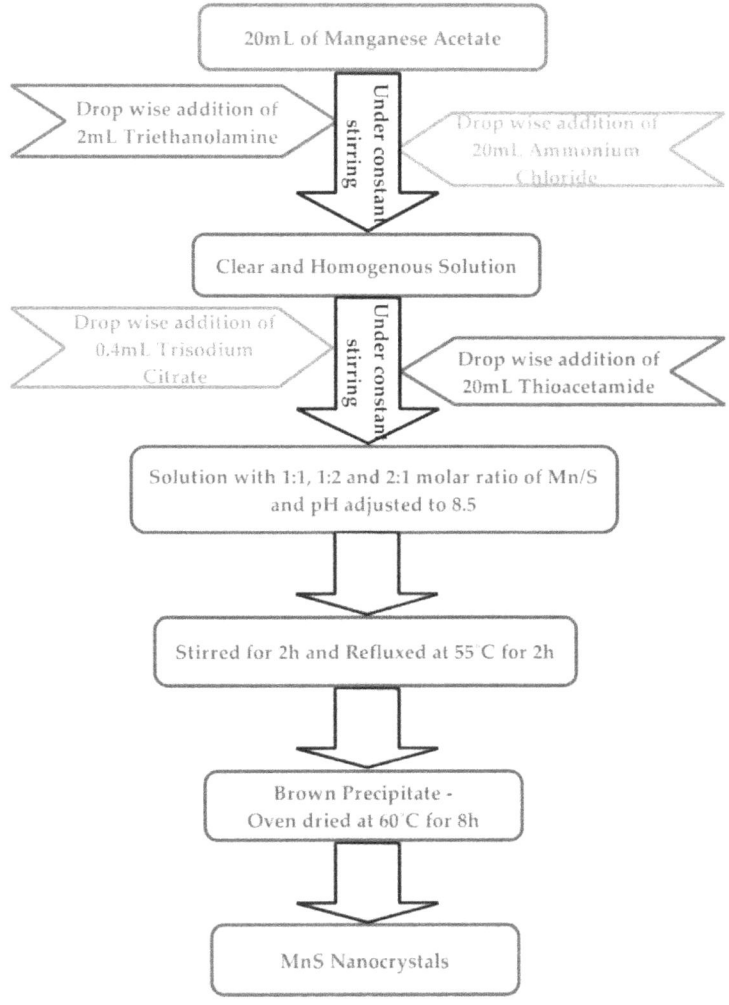

Fig. 5.1: A schematic diagram of the formation of MnS nanocrystals

A typical MnS nanocrystals were synthesized by wet chemical route is described as follows: 20mL of aqueous solutions with 1:1, 1:2 and 2:1 molar ratio of manganese acetate to thioacetamide were used as manganese source and sulphur source respectively.

2mL of triethanolamine was mixed with an aqueous solution of 20mL manganese acetate slowly. This solution was stirred for few minutes. Under the continuous stirring, 20mL of ammonium chloride was added drop-wise, the solution became clear and homogenous. The mixture was stirred for several minutes, 0.4mL trisodium citrate was added in, followed by addition of 20mL of thioacetamide solution. The pH of final solution was adjusted to 8.5 and stirred for 2h. All the above process was performed at the room temperature. The reaction mixture was refluxed at 55°C for 2h in water bath. After the reaction, the brown precipitates were separated from the solution by centrifugation, washed thoroughly several times with deionized water and oven dried at 60°C for 8h.

5.3 Results and Discussion

5.3.1 Structural Properties

In order to determine the size and to study the structural properties of the synthesized MnS nanocrystals, the powder XRD analysis has been performed. Scanning electron microscope (SEM) is used for the morphological study of MnS nanocrystals. Crystallite size and morphology play important roles in the applications, which drive the modern studies to focus on the synthesis of new nanocrystalline materials. Functional groups present in the nanocrystals are found using Fourier transform infrared spectroscopy (FTIR).

5.3.1.1 XRD Analysis

The crystalline phase and orientation of the as-prepared MnS nanocrystals at different precursor molar ratio are studied by X-ray diffraction pattern. The sharp and high intensity diffraction peaks imply good crystallinity of the samples. Fig. 5.2(a) represents the XRD pattern for the sample with 1:1 molar ratio of Mn and sulfur source. The peaks are observed at 2θ values of 27.5°, 45.5° and 54.7°.

The diffraction peaks can be indexed to the metastable face-centered sphalerite structured β-MnS with lattice constants a=b=c=5.611A° which is in good agreement with the standard data from JCPDS card 40-1288 [8]. Fig. 5.2(b) shows the X-ray diffraction pattern of the as-prepared sample with 1:2 molar ratio of Mn and sulfur source. It exhibits prominent peaks at 26.32°, 28.00° and 30.41° with the lattice parameter values of a=b=3.962A° and c=6.252A°. Fig. 5.2(c) shows the XRD pattern of the sample with 2:1 molar ratio of Mn and sulfur source. The peaks are observed at 2θ values of 26.14°, 27.80° and 29.88° with the lattice constants of a=b=3.95A° and c=6.262A°. The diffraction peaks of samples with 1:2 and 2:1 molar ratio are in well agreement with the standard JCPDS card 40-1289 [9] which shows the formation of metastable γ-MnS wurtzite structure. In the XRD pattern, there is no evidence of superfluous peaks of $Mn(OH)_2$ and MnO.

In order to reduce the errors in crystallite size calculation, the modified Scherrer formula is used in that all the diffraction peaks or any number of selected peaks is used to obtain the average crystallite size by least squares method. The crystallite size is estimated from XRD data by modified Scherrer equation [10] as follows, taking logarithm on both sides of basic Scherrer formula

$$In\beta_r = In(C\lambda/D.cos\theta) = In(C\lambda/D) + In(1/cos\theta) \quad (5.1)$$

The linear regression plot is obtained from the results of $In(\beta_r)$ against $In(1/cos\theta)$ for the MnS nanocrystals as shown in Fig 5.3 (a-c). The single straight line through all the points gives an intercept In Cλ/D which can be used to calculate the crystallite size of the samples. The average crystallite size (D) is found to be 2.04 nm, 5.56 nm and 1.74 nm for the MnS sample with 1:1, 1:2 and 2:1 molar ratio respectively. The crystallite size of prepared MnS nanocrystals strongly increases with increase in sulfur molar concentration. On the other hand, the crystallite size decreases with increase in manganese molar concentration.

The change in crystallite size is due to the change in molar concentration of precursors is shown in Table 5.1. The microstrain (ε) developed in the nanocrystals can be calculated from the relation [11],

$$\epsilon = (¼)\beta \cos\theta \tag{5.2}$$

It is observed that the micro strain decrease with increase in crystallite size, as shown in Table 5.1, which indicates a lower number of lattice imperfections.

Table 5.1: XRD data analysis for Crystallite size, Micro strain and Dislocation density of the MnS nanocrystals with 1:1, 1:2 and 2:1 molar ratio

Prepared Samples	2θ	β_o = FWHM in radians	$\beta_r^2 = \beta_o^2 - \beta_i^2$	$\ln(\beta_r)$	$\ln(1/\cos\theta)$	Crystallite size (D) in nm	Micro strain (ε)	Dislocation density (δ)
1:1 molar ratio	27.50	0.0565	0.0031	-2.87	1.029	2.04	0.0137	0.2403
	45.50	0.0346	0.0012	-3.36	1.084		0.0079	
	54.70	0.0295	0.0008	-3.52	1.126		0.0065	
1:2 molar ratio	26.32	0.0249	0.0006	-3.69	1.027	5.56	0.0060	0.0323
	28.00	0.0249	0.0006	-3.69	1.031		0.0060	
	30.41	0.0249	0.0006	-3.69	1.036		0.0060	
2:1 molar ratio	26.14	0.0561	0.0031	-2.87	1.027	1.74	0.0136	0.3303
	27.80	0.0523	0.0027	-2.95	1.030		0.0126	
	29.88	0.0502	0.0025	-2.99	1.035		0.0121	

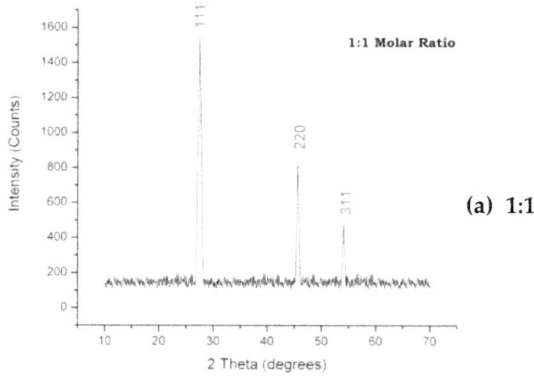

(a) 1:1

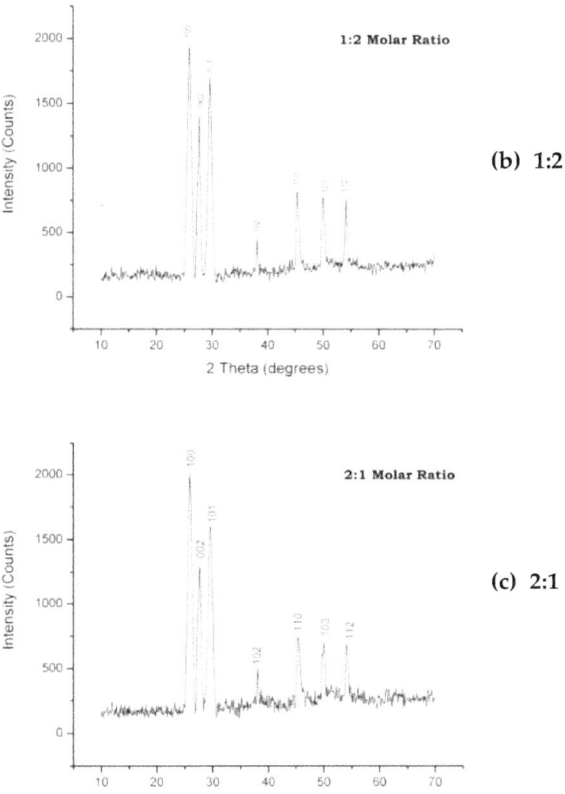

Fig. 5.2: XRD Patterns of MnS Nanocrystals with Different Precursor Molar Ratios (a) 1:1, (b) 1:2 and (C) 2:1

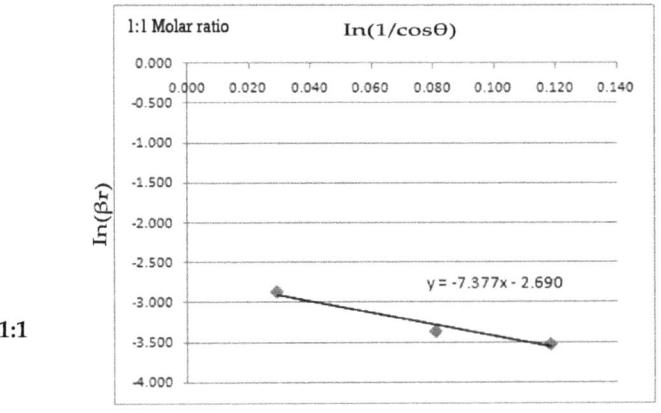

(b) 1:2

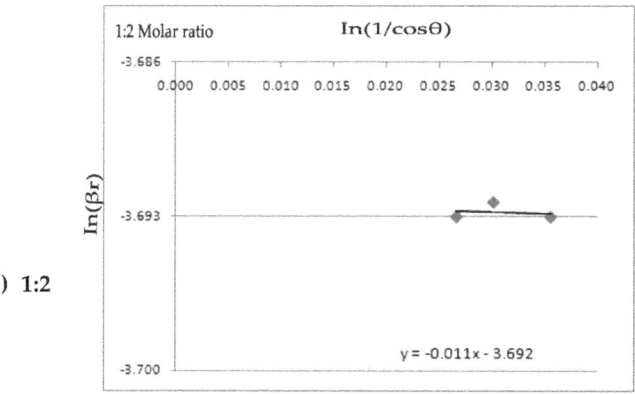

(c) 2:1

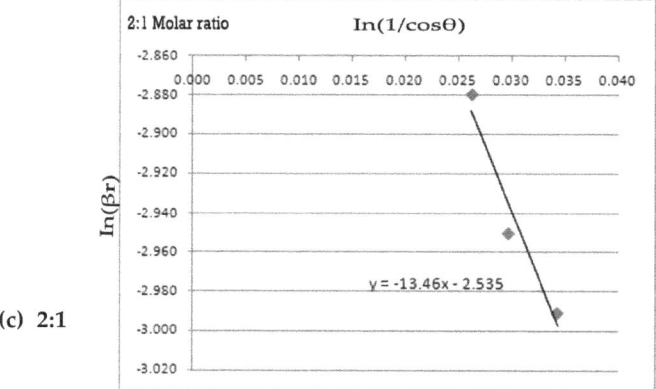

Fig. 5.3: Linear plots of modified Scherrer equation and obtained intercepts for the MnS nanocrystals with different precursor molar ratios (a) 1:1, (b) 1:2 and (C) 2:1

The dislocations are an imperfection in a crystal associated with misregistry of the lattice in one part of the crystal with respect to another part. Unlike vacancies and interstitial atoms, dislocations are not equilibrium imperfections. In fact growth mechanism involving dislocation is a matter of importance. The dislocation density of the sample is given by the relation,

$$\delta = n/D^2 \qquad (5.3)$$

where n is a factor, which equals unity giving minimum dislocation density and D is the crystallite size. The dislocation densities of the MnS sample are listed in Table 5.1. There is an increase in crystallite size corresponding to decrease in dislocation density are observed in the as-synthesized MnS nanocrystals.

5.3.1.2 SEM Analysis

The surface morphology of synthesized nanocrystals has been studied by using scanning electron microscope. The SEM image of the products obtained with precursor molar ratio of 1:1, 1:2, and 2:1 are shown in Fig. 5.4 (a-c). The accelerating voltage, magnification, spot size and instrumental parameters are indicated on SEM images. The surface morphology can be found from the analysis of the obtained SEM images at different magnification. The SEM indicates that the appearances of as-prepared MnS particles are spherical in shape.

In Fig. 5.4(a), the rough surface spherical MnS particles are observed in the sample prepared with 1:1 molar ratio of the Mn and sulfur source. The diameter of the spheres varies within 1-3 μm. Fig. 5.4(b) shows the SEM image of the sample prepared at the molar ratio of 1:2 which reveals the smooth spherical crystals, with an average diameter of about 3-5 μm. The sample produced with excess of sulfur source has leads to increase the particle size. Fig. 5.4(c) shows the SEM image of the sample synthesized with precursor molar ratio of 2:1 [Mn^{2+} and S^{2-}]. The spherical clusters of MnS particles are formed in the morphology of this sample due to excess of Mn source. The average diameter of spheres is about 500 nm to 1 μm along with few undefined structures. The MnS particles able to agglomerated because of their high surface energy and high surface tension.

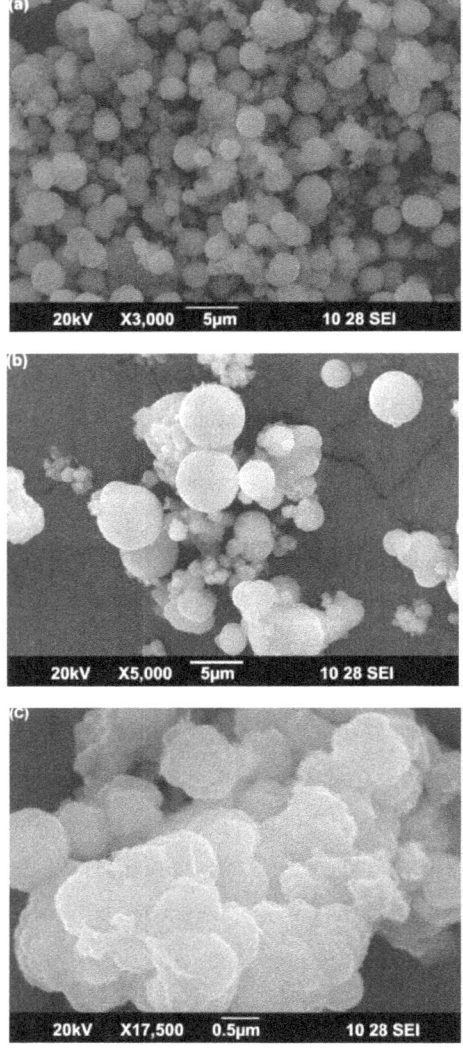

Fig. 5.4: SEM images of the sample synthesized with different precursor molar ratios (a) 1:1, (b) 1:2 and (c) 2:1

5.3.1.3 FTIR Spectra Analysis

FT-IR is one of the most general spectroscopic techniques used to identify the functional groups in materials. It is important and popular tool exposition and compound identification.

Fig. 5.5 (a-c) shows the FT-IR spectra of the as-prepared MnS nanocrystals with 1:1, 1:2 and 2:1 molar ratio in the frequency region from 400 cm^{-1} to 4000 cm^{-1}. A peak observed around 415 to 425 cm^{-1} is assigned to Mn-S bending [12]. The band between 450 and 600 cm^{-1} is attributed to S-S stretching. The symmetric and asymmetric C-S stretching vibrations are indicated in the range of 600-800 cm^{-1}. The relative intense S-O stretching absorption bands in the 800 cm^{-1} to 1000 cm^{-1} region of the IR spectrum. The C-O stretching vibrations (acetate) at 1016 & 1072 cm^{-1}, 1026 & 1066 cm^{-1}, and 1022 & 1070 cm^{-1} are obtained [13]. The C-N stretching absorption occurs in the region 1257 cm^{-1}, 1244 cm^{-1} and 1247 cm^{-1}. The band at 1394cm^{-1}, 1398cm^{-1} and 1408cm^{-1} confirms the presence of capping agent trisodium citrate [14].

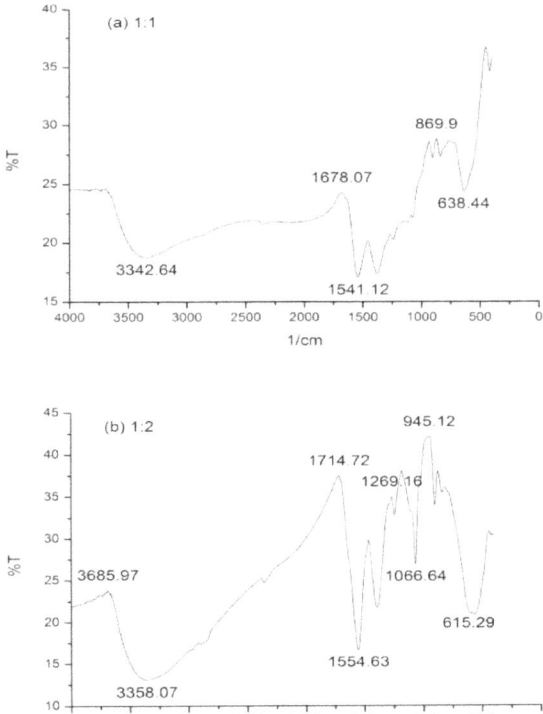

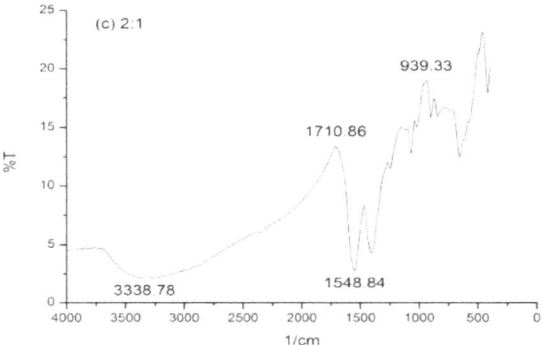

Fig. 5.5: FTIR spectra of the MnS nanocrystals with different precursor molar ratios (a) 1:1, (b) 1:2 and (C) 2:1

A broad peak centered around 3375 cm^{-1}, 3365 cm^{-1} and 3317 cm^{-1} can be assigned to the O-H stretching and a band at 1558 cm^{-1}, 1554 cm^{-1} and 1548 cm^{-1} which can be assigned to the O-H bending modes of water [15]. The peaks observed at 2347 cm^{-1} and 2358 cm^{-1} correspond to C-H asymmetrical stretching [16]. The typical bands in 3000-2800 cm^{-1} range due to -C-H symmetric and asymmetric stretching vibrations modes of alkyl chains experience negligible shifts.

5.3.2 Optical Properties

Optical properties of the MnS nanocrystals have been investigated by UV-Vis absorption and photoluminescence (PL) spectroscopy.

5.3.2.1 UV-Visible Absorption Spectra Analysis

The UV-Visible optical absorption spectra of MnS nanocrystals are studied and it is shown in Fig. 5.6. The 0.001g samples are ultrasonically dispersed in 1ml absolute ethanol. UV-Visible absorbance spectra indicate that each sample having two absorbance peaks about 206-232 nm and 263-265 nm. The broad absorption extended up to 800 nm. It shows a wide range of absorption from UV to NIR which indicates that the prepared material is good for sun light absorption.

The absorption edges (electronic transitions) of as-prepared MnS samples with 1:1, 1:2 and 2:1 molar ratio are 220 & 264 nm, 232 & 265 nm, and 206 & 263 nm respectively. The absorbance spectra show the large blue shift which is compared with the bulk value of 326 nm at 0K [17, 18]. The large blue shift is concerned with a quantum size effect which is due to the confinement of electrons and holes in a small volume [19]. The peak of the spectra corresponds to the fundamental absorption edges in the samples, and could be used to estimate the optical band gap of the nanocrystals [20]. Due to quantum size effect, the band gap energy increases with decreasing particle size. The optical energy band gap (E_g) of MnS nanocrystals is calculated using the relation,

$$E_g = hc/\lambda \; (eV) \tag{5.4}$$

where, h is Plank's constant, c is speed of light and λ is cut off wavelength.

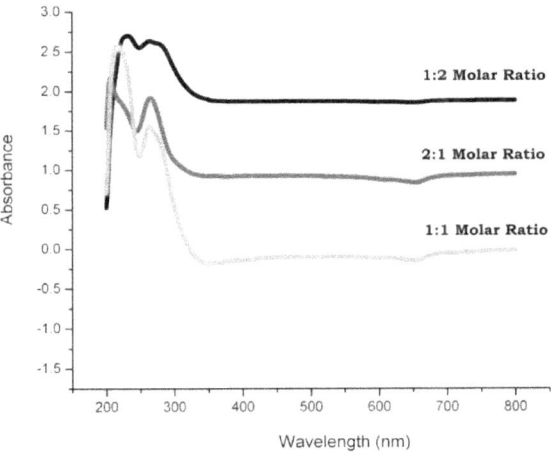

Fig. 5.6: UV-Visible absorption spectra of the MnS nanocrystals with 1:1, 1:2 and 2:1 molar ratio

The direct optical band gap values of the MnS samples with 1:1, 1:2 and 2:1 molar ratio are 4.71 eV, 4.69 eV and 4.72 eV respectively. A subsequent decrease in absorbance at higher concentration of Mn^{2+} may be attributed to shadow effect, where the absorbing molecules are effectively screened from the incident radiation by presence of large quantity of respective molecules in its path. The MnS nanocrystals are wide band gap semiconductor which can be used as optical window in optoelectronic devices [21].

5.3.2.2 PL Spectra Analysis

The photoluminescence spectra measured at room temperature of as-synthesized MnS nanocrystals is shown in Fig. 5.7. The excitation wave length is set to as 350 nm for all the prepared samples. The PL spectra revealed that the two emission peaks centered at 420 & 441 nm, 409 & 433 nm, and 424 & 442 nm for the MnS samples with 1:1, 1:2 and 2:1 molar ratio respectively. The as-prepared MnS nanocrystals exhibit a strong blue emission (λ=400–450 nm). The width of peak intensity increases as the crystallite size of the particle decreases [22]. Broadening of the emission shoulder is attributed to both size distribution and increase in the surface states owing to the increase in surface to volume ratio for nanocrystals. With increase in molar concentration of Mn (2:1), the PL emission peaks are found to be blue shifted whereas red shift is observed in the emission band when increasing molar concentration of sulfur (1:2).

The blue emission peaks are assigned to the energy level of sulfur vacancies with the holes from the valance band and interstitial Mn from interstitial sulfur. Although, the emission peak at around 420 nm is assigned to deep trap emissions or defect related emission of MnS [23]. The peak at around 440 nm may be attributed to the presence of various surface states. In addition to that the benefits of blue emission are in the light emitting devices and biological fluorescence labeling applications.

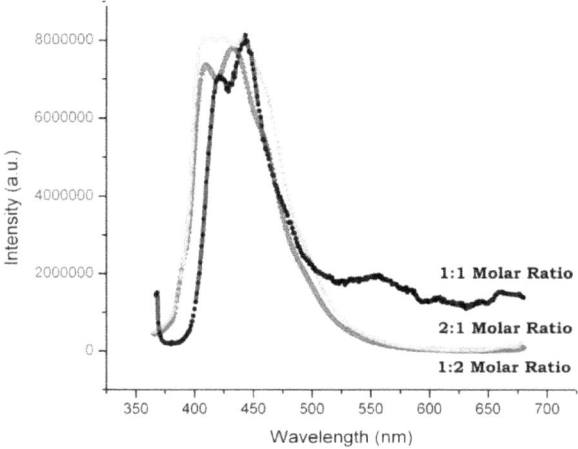

Fig. 5.7: PL spectra of the MnS nanocrystals with 1:1, 1:2 and 2:1 molar ratio

5.3.3 Electrical Properties

The LCR meter data for the MnS samples with 1:1, 1:2 and 2:1 molar ratio are used to evaluate different electrical parameters like dielectric constant (ϵ_r), dielectric loss (tan δ), impedance (Z), Electric modulus, Relaxation time (τ), AC conductivity (σ_{ac}) and activation energy (E_a). The characterization of dielectric behavior is very important to understand the polarization mechanism and from application point of view, the temperature and the frequency dependence of dielectric behaviors are very important. The relative dielectric constant of the material determines its ability to store electrostatic energy. The loss tangent indicates the ability of dielectrics to support the electrostatic field, while dissipating minimal energy in the form of heat. The dielectric dispersion behavior offers an opportunity to gain vital insight into the details of ionic conduction processes. The electrical conductivity studies indicate the nature of dominant charge species involved in the conduction on application of external electric field. Similarly, the electrical impedance formalism helps to understand the relaxation of defect species.

5.3.3.1 Dielectric Studies

The dielectric properties of materials are mainly due to contributions from the electronic, ionic, dipolar and space charge polarizations. The measurement of dielectric parameters like the dielectric constant and loss as a function of frequency and as a function of different temperatures reveals the electrical processes that take place in MnS nanocrystals. The variations of the dielectric constant and dielectric loss of the MnS nanocrystals with respect to the frequencies of 50Hz to 5MHz at different temperatures of 323K to 473K are displayed in Fig. 5.8(a-f).

The dielectric constant and dielectric loss decreases in all the three samples (1:1, 1:2 and 2:1) with increase in frequency and reaches a constant value at high frequencies. It could be due to the dipoles not being able to follow the field variation at high frequencies and also due to the polarization effects [24]. At high frequencies the periodic reversal of the electric field occurs so fast that there is no excess ion diffusion in the direction of the field.

The variation of dielectric constant and dielectric loss with respect to temperature at various frequencies are depicted in Fig 5.9(a-f). It is observed that when the temperature rises, dipolar rotations and molecular motions become easier and the response to field variation is reinforced. Therefore, both the dielectric constant and the loss factor increase with temperature. This is justified from the effect of temperature, as the temperature facilitates the diffusion of ions in the space charge polarization [25], consequently the value of dielectric constant and dielectric loss increases with temperature. The high temperature values of dielectric constant as well as that of dielectric loss increase with decreasing frequency. This increase in dielectric response with temperature may be due to interfacial polarization dominating over dipolar polarization.

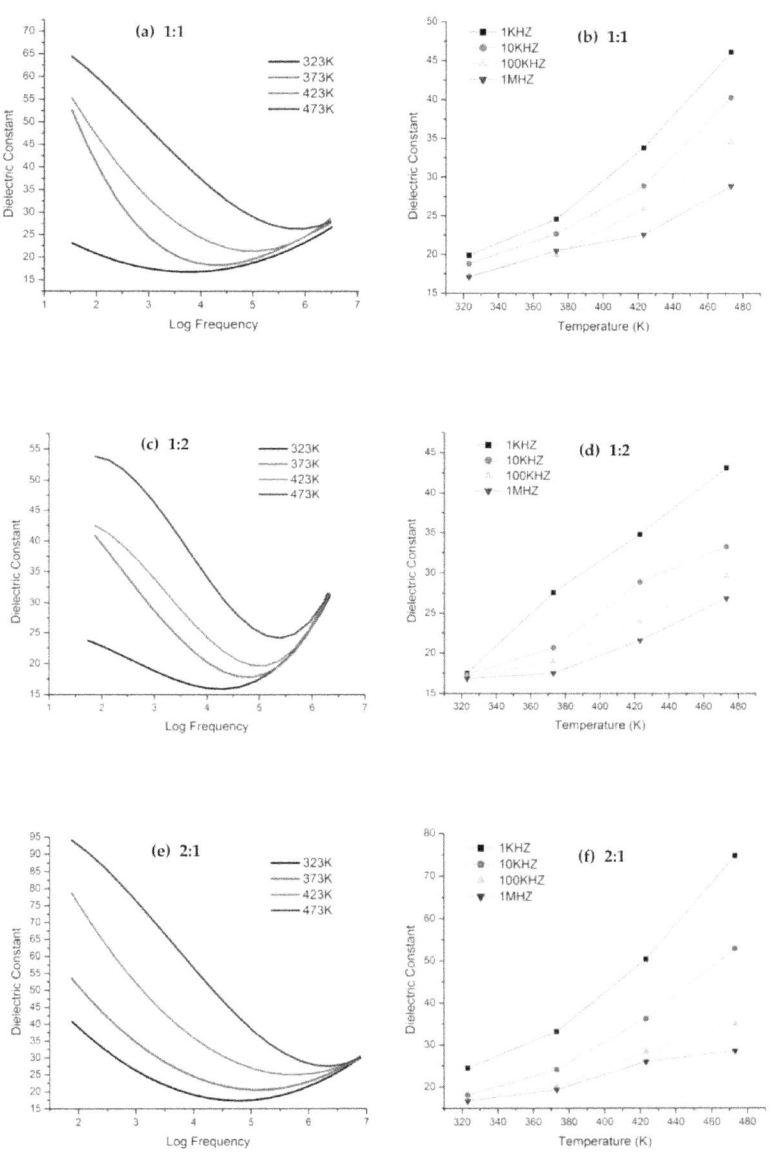

Fig. 5.8: Frequency and temperature dependence of dielectric constant for the MnS nanocrystals with different precursor molar ratios (a & b) 1:1, (c & d) 1:2 and (E & F) 2:1

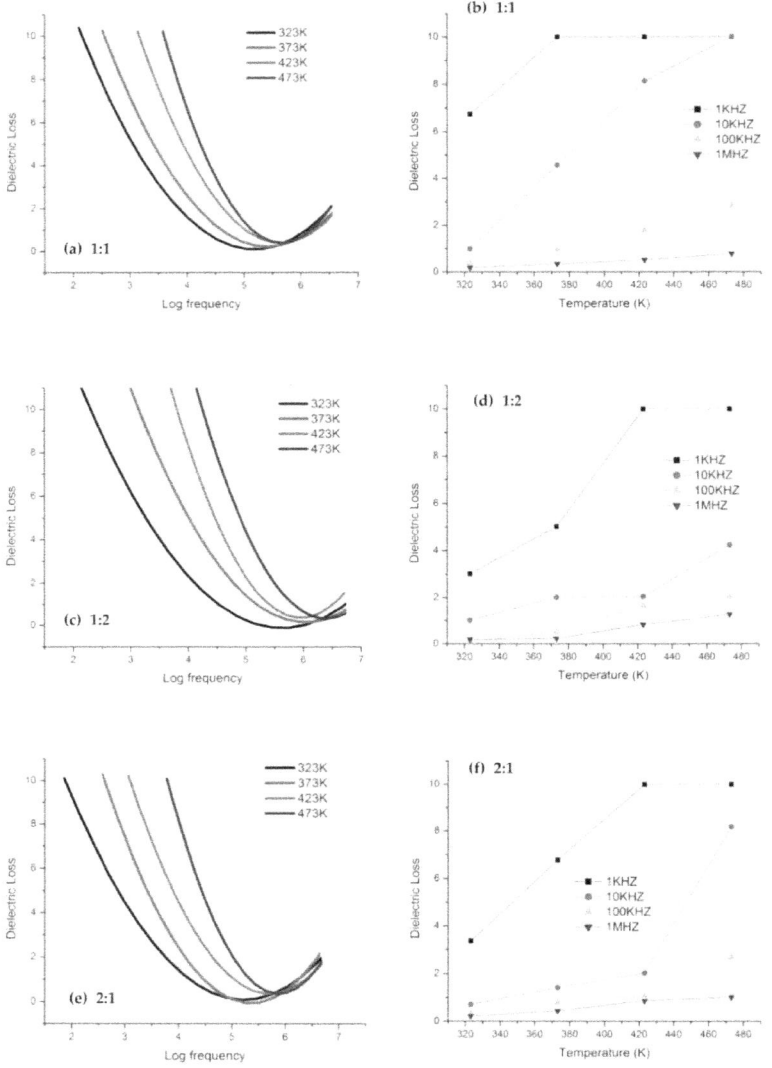

Fig. 5.9: Frequency and temperature dependence of dielectric loss for the MnS nanocrystals with different precursor molar ratios (a & b) 1:1, (c & d) 1:2 and (E & F) 2:1

It indicates that the onset of some additional relaxation mechanism in the material attributed to the AC conduction mechanism being dominant at high temperature [26].

For all MnS samples, contribution of the reorientation of the off-centre ions coupling with the thermally activated conduction electrons may appear due to ionization of the sulfur vacancies and results into such response [27].

5.3.3.2 Impedance Studies

Fig. 5.10(a-c) depicts the experimental values of |Z| versus frequency at different temperatures of the applied AC field for the MnS samples with 1:1, 1:2 and 2:1 molar ratio. It can be seen that |Z| considerably decreases as the frequency increases. This indicates that the components of capacity and resistance of the equivalent circuit are active in this range of frequencies. However, for the frequencies less than 10KHZ, |Z| decreases as temperature increases, implying a decrease in the total resistance of the MnS sample. In other words, the value of impedance remains constant in the temperature range 323-473K at the highest frequency above 100KHZ. The |Z| values for all temperatures merging at high frequency may be due to the release of space charges, as a result of reduction in barrier properties of the material with rise in temperature [28, 29]. In complex impedance spectra or Cole–Cole plots, Z' values represents the resistive part and the Z'' values represents the capacitive part.

When more than one dielectric relaxation is involved, this complex representation of the dielectric data is used to evaluate the resistance (R) and capacitance (C) values associated with grain and grain boundary of the material [30]. Fig. 5.11(a-c) shows the impedance Cole-Cole plot of nanocrystalline MnS for three different precursor molar ratios of 1:1, 1:2 and 2:1 at four different temperatures (323K, 373K, 423K and 473K). There are traces for two semicircles, one at the low frequency region and the other at the high frequency region (not visible). A small semicircle at high frequencies indicates the effect of grain and large semicircle at low frequencies indicates the grain boundary effect [31].

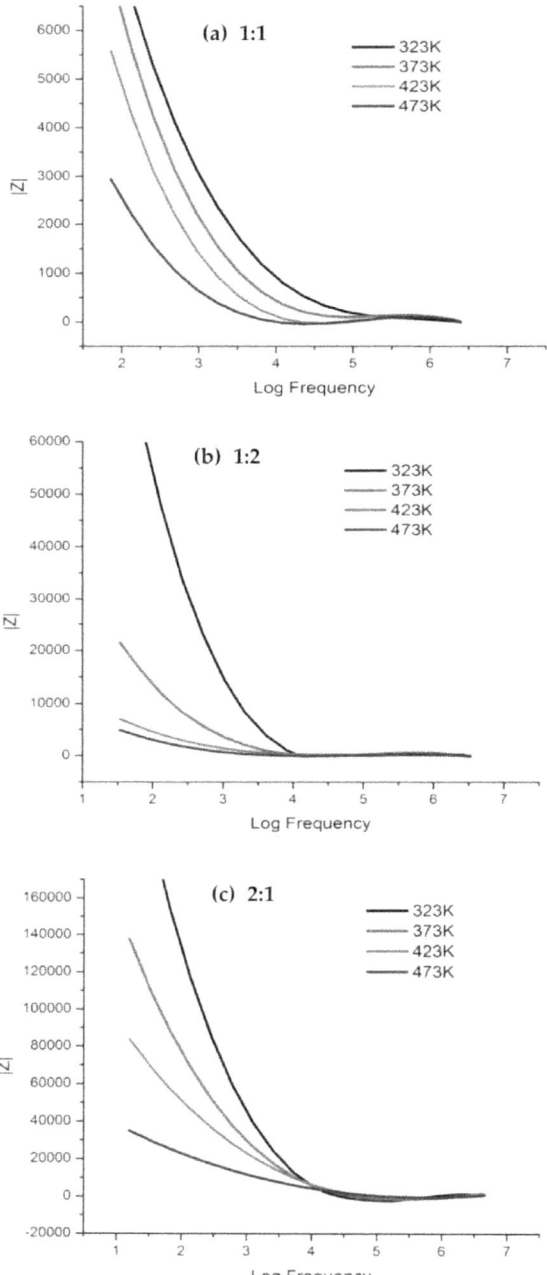

Fig. 5.10: Frequency dependence of |Z| at different temperatures for the MnS nanocrystals with different precursor molar ratios (a) 1:1, (b) 1:2 and (C) 2:1

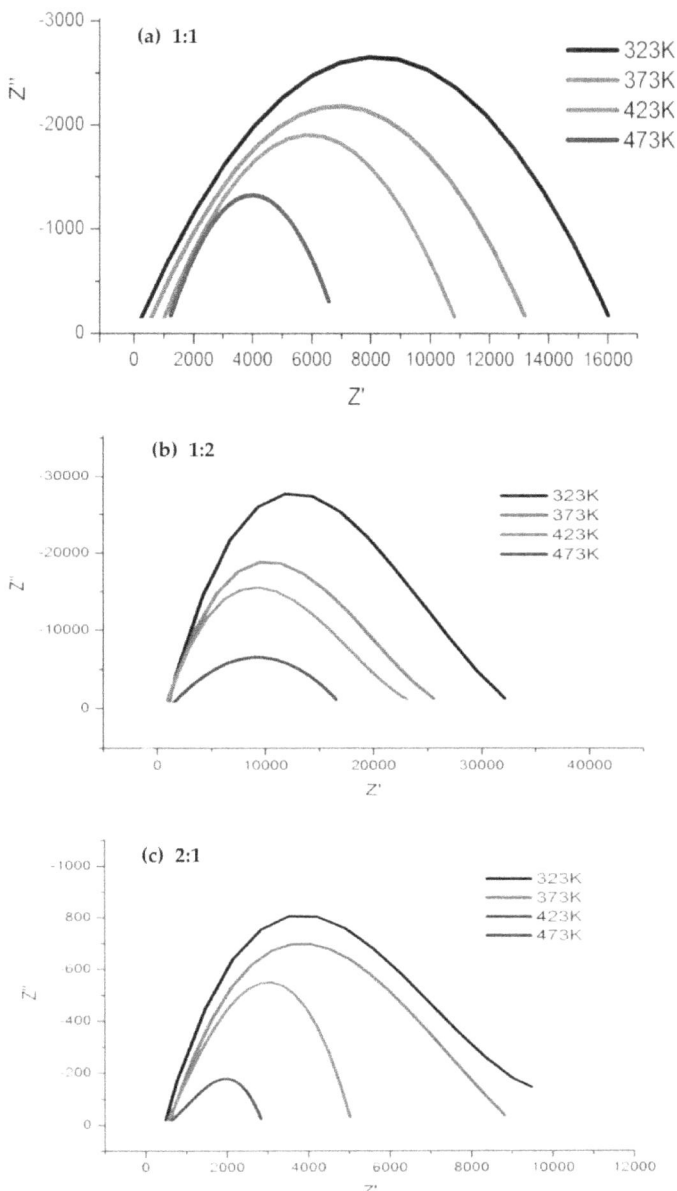

Fig. 5.11: Complex impedance plots of the MnS nanocrystals with different precursor molar ratios (a) 1:1, (b) 1:2 and (C) 2:1

The intercept of the semicircles on the real axis gives the resistance of grain (R_g) and grain boundary (R_{gb}) of the corresponding component contributing towards the impedance of the samples. The impedance of nanostructured materials depends on both grain size and temperature as in evident from the plots. The size of the Cole-Cole plots decreases with increase in temperature. The capacitance (C) of each component can be calculated using the formula [32]

$$\omega_{max}\tau = 2\pi f_{max} RC = 1 \tag{5.5}$$

f_{max} is the frequency of the maximum of the semicircle, and τ is the relaxation time.

The various grain boundary parameters are observed and calculated values are tabulated in Table 5.2. The circuit consists of a series array of two sub circuits, one represents grain effect and the other represents a grain boundary effect which is shown in Fig. 5.12. Each sub circuit is composed of a resistor and capacitor joined in parallel. Let (R_g, R_{gb}) and (C_g, C_{gb}) be the resistances and capacitances of grains and grain boundaries, respectively.

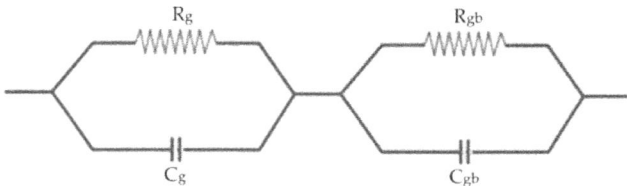

Fig. 5.12: Equivalent circuit of the as-prepared MnS samples from the impedance spectra

5.3.3.3 Electric Modulus Studies

The Z' and Z'' are used for the evaluation of real (M') and imaginary (M'') parts of complex electrical modulus (M*). The frequency dependence of dielectric modulus (M' and M'') over a wide range of frequency at different temperatures is studied.

The electric modulus corresponds to the relaxation of the electric field in the material when the electric displacement remains constant. The electric modulus represents the real dielectric relaxation process, which can be expressed as

$$M^*(\omega) = \frac{1}{\varepsilon^*(\omega)} = M' + M'' = M_\infty \left[1 - \int_0^\infty \left(-\frac{d\phi(t)}{dt} \right) \exp(-i\omega t) dt \right] \quad (5.6)$$

where $M_\infty = (\varepsilon_\infty)^{-1}$ is the asymptotic value of $M'(\omega)$ and $\phi(t)$ is the time evolution of the electric field within the material [33, 34]. The frequency dependence of the electric modulus M' and M'' for various temperatures is shown in Fig. 5.13(a-f) for the MnS sample with 1:1, 1:2 and 2:1 molar ratio.

Table 5.2: Values of the equivalent circuit parameters deduced from the impedance spectra for the MnS nanocrystals with 1:1, 1:2 and 2:1 molar ratio

Samples and Factors		Grain resistance (R_g) Ω	Grain boundary resistance (R_{gb}) Ω	Grain capacitance (C_g) F	Grain boundary capacitance (C_{gb}) F
1:1 molar ratio	323K	248	16000	2.29E-07	3.55E-09
	373K	242	13236	8.33E-08	1.52E-09
	423K	213	10879	2.64E-08	5.17E-10
	473K	199	6668	2.01E-08	5.99E-10
1:2 molar ratio	323K	532	33639	4.73E-07	7.48E-09
	373K	435	26931	2.30E-07	3.72E-09
	423K	412	23828	1.37E-07	2.36E-09
	473K	403	17333	9.90E-08	2.30E-09
2:1 molar ratio	323K	180	9611	1.06E-07	1.98E-09
	373K	166	8925	3.80E-08	7.07E-10
	423K	138	5720	1.29E-08	3.11E-10
	473K	110	2837	4.56E-09	1.77E-10

It is evident from Fig. 5.13(a-f) that for all temperatures, M' approaches nearly zero at lower frequencies which confirming the presence of an appreciable electrode and/or ionic polarization in the temperature range. M' reaches a constant value at higher frequencies.

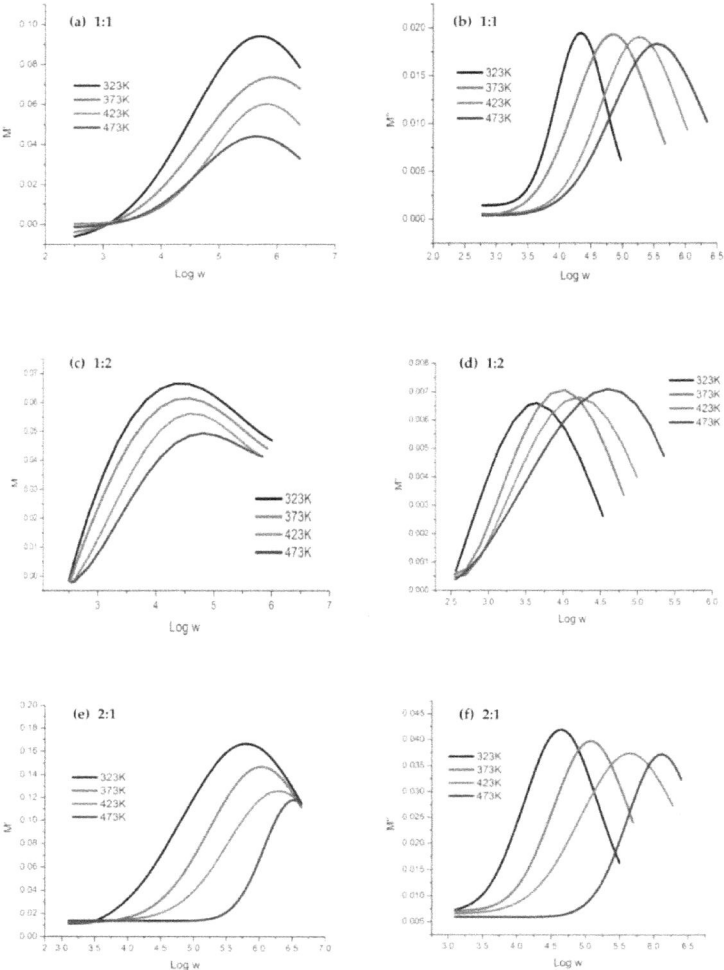

Fig. 5.13: Frequency dependence of M' and M'' at different temperature for the MnS nanocrystals with different precursor molar ratios (a & b) 1:1, (c & d) 1:2 and (E & F) 2:1

The value of M' increases from the low frequency value towards the high frequency value as temperature increases. M' shows a dispersion tending toward M_∞, while M'' exhibits a maximum M''_{max} which shifts to higher frequencies with increase in temperature. The frequency region below peak maximum M'' determines the range in which charge carriers are mobile on long distances.

At frequency above peak maximum M″, the carriers are confined to potential wells, being mobile on short distances [35]. The frequency ω_{max} gives the relaxation time (τ) from the condition $\omega_{max}\tau=1$ [36]. The relaxation time versus temperature behavior observed that the relaxation time degreases with the increase in temperature, indicating that the relaxation phenomenon is thermally activated.

5.3.3.4 AC Conductivity Studies

The AC conductivity with different frequencies at different temperatures is determined by using the following relation

$$\sigma_{ac} = \omega \varepsilon_0 \varepsilon' \tan \delta \qquad (5.7)$$

where ε_0 is the permittivity of free space (8.854 x 10^{-12} F/m) and ω is the angular frequency ($\omega = 2\pi f$). The general trend of conductivity is to increase according to increase in temperature. It can be observed in Fig. 5.14(a, c & e) that AC conductivity increases with the increase in frequency but at higher frequency the conductivity tend to constant for all temperatures. If the frequency of AC applied field increases, the hopping frequency of electron increases which causes an increase in the mobility of charge carriers. Thus, it observed that a gradual increase in conductivity with frequency. At higher frequencies, the hopping of electron cannot follow the applied field frequency so that the conductivity is almost constant for all temperatures. The increase in conductivity does not represent the increase in charge concentration but increase in mobility of charge carriers [37].

The variations of AC conductivity with temperature at preferred frequencies for the MnS samples are shown in Fig. 5.14 (b, d & f). The AC conductivity decreases with inverse temperature for all frequencies, the decrease being more rapid at lower frequencies. This decrease in the conductivity with temperature is due to the decrease in drift mobility of the charge carrier thereby reducing the charge hopping.

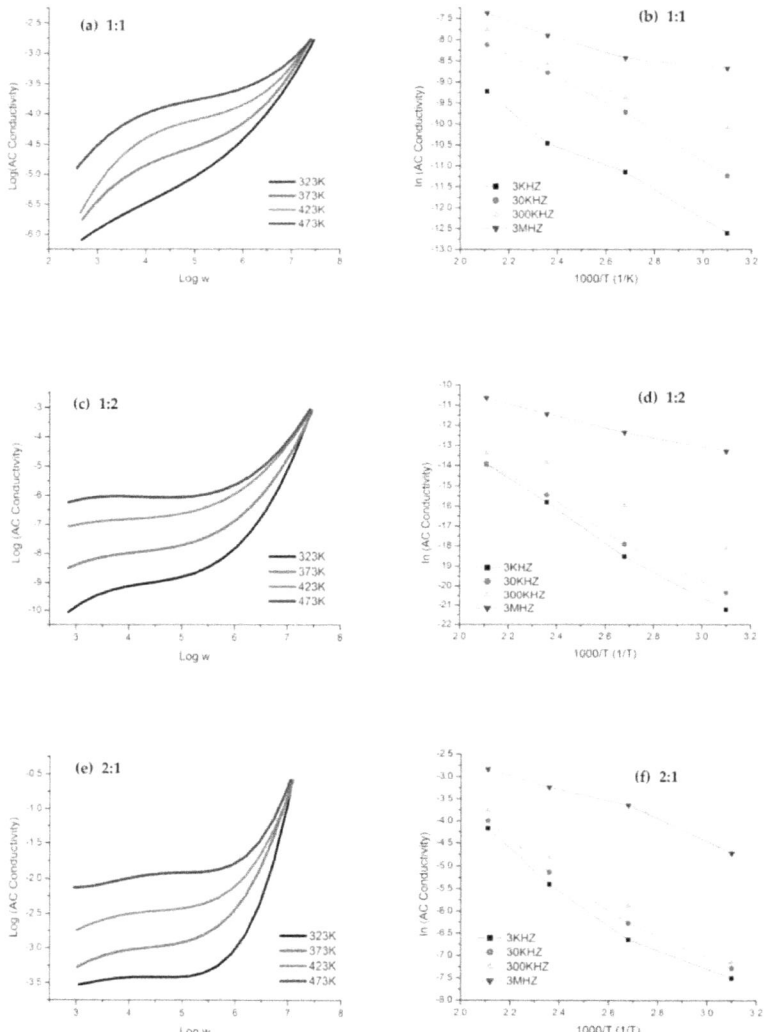

Fig. 5.14: Frequency and temperature dependence of AC conductivity for the MnS nanocrystals with different precursor molar ratios (a & b) 1:1, (c & d) 1:2 and (E & F) 2:1

The AC conductivity (σ_{ac}) is thermally activated processes which can be used to calculate the activation energy (E_a). The behavior of the activation energy describes the temperature dependence of σ_{ac}. In general, the activation energy is related to the conductivity by the Arrhenius relation,

$$\sigma_{ac} = \sigma_0 \exp(-qE_a/kT) \qquad (5.8)$$

where, σ_0 is the pre-exponential factor, k is the Boltzmann constant and T is the temperature. The values activation energy obtained from the slope of Ln (σ_{ac}) vs q/kT plot for each frequency. The variation of E_a with frequency of as-synthesized MnS nanocrystals is shown in Fig. 5.15. The activation energy is found to be 0.35 eV (1:1), 0.64 eV (1:2) and 0.28 eV (2:1) at 3 KHz. The value of activation energy decreases with the increase of frequency confirms that hopping conduction is the dominant current transport mechanisms. Thus, the increase of the applied frequency enhances the electronic jumps between the localized states; consequently, the activation energy, decreases with increasing frequency [38, 39].

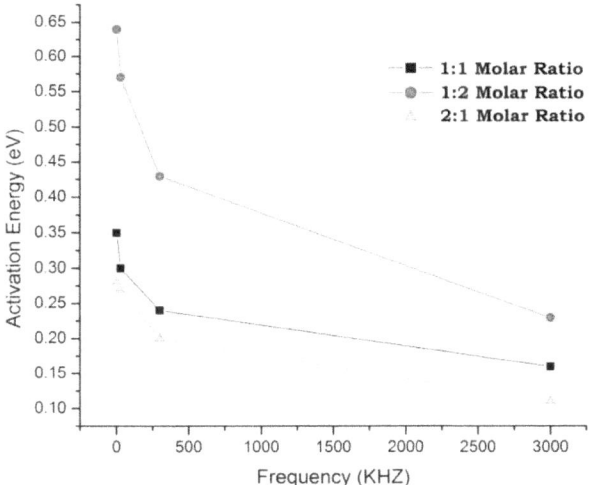

Fig. 5.15: Variations in activation energy with frequency for the as-prepared MnS nanocrystals

5.3.4 Magnetic Properties

The variation of magnetic moment as function of magnetic field for the as-synthesized MnS nanocrystals at RT using vibrating samples magnetometer is shown in Fig. 5.16. The coercive field of the samples approximately 2.31, 8.42, and 5.57 Oe and the saturation magnetization [Ms] of the samples are found to be 0.56, 0.38, 0.57 emu/g for the MnS samples with 1:1, 1:2, and 2:1 molar concentration, respectively which is shown in Table 5.3. The samples having a small remanence and coercivity is said to be magnetically soft with paramagnetic behavior [40, 41]. The squareness ratios (Mr/Ms) are found to be 0.00016, 0.00059 and 0.00038 for the MnS samples prepared with 1:1, 1:2 and 2:1 molar ratio respectively. It suggests that the samples having superparamagnetic grains with multi domain grains. The incompletely filled inner electron shells of Mn^{2+} ions are generally responsible for the paramagnetic behavior of samples and they have unpaired electrons with free spin magnetic moments.

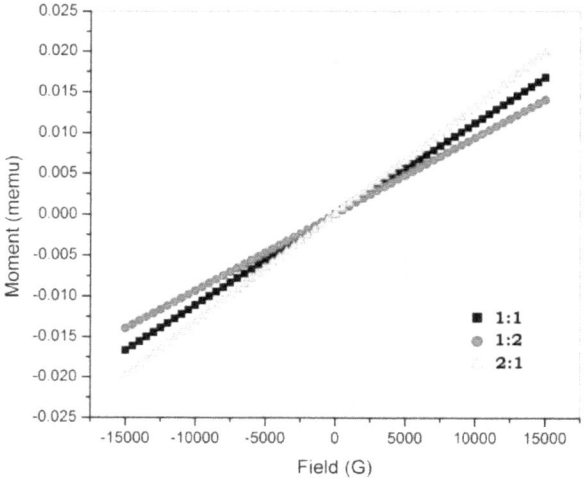

Fig. 5.16: Magnetic moment vs. magnetic field plots of the MnS nanocrystals with 1:1, 1:2 and 2:1 molar ratio

Table 5.3: Magnetic properties of the MnS nanocrystals with 1:1, 1:2 and 2:1 molar ratio

Samples	Ms (emu/g)	Mr (memu/g)	Mci (Oe)	Effective magnetic moment (μB)
1:1 molar ratio	5.65E-01	8.89E-05	2.317	1.24
1:2 molar ratio	3.89E-01	2.28E-04	8.425	1.01
2:1 molar ratio	5.76E-01	2.22E-04	5.578	1.21

The magnetic susceptibility is calculated from the slop of the magnetization versus applied field [42]. The susceptibility for the MnS samples with 1:1, 1:2, and 2:1 molar ratio has estimated as 4.01×10^{-5}, 2.06×10^{-5}, and 3.85×10^{-5} respectively and its relative permeability is only slightly greater than one and it is independent of magnetic field strength. Multi-domains need less magnetic field to switch compared with the single domain state [43]. Accordingly, it is found that at large crystallite size, the switching field decreases and the magnetization saturation increases compared with the smaller size. The effective magnetic moment per Mn^{2+} ion estimated from

$$\mu_{eff} = \sqrt{3kT\chi_m / N_A} \qquad (5.9)$$

It is found to be 1.24 µB, 1.01 µB and 1.21 µB for the MnS samples with 1:1, 1:2 and 2:1 molar ratio respectively. The effective magnetic moment is less than the theoretical value of 5.91 μ_B, which might be related to a strong spin-orbit coupling in the sample. It may be due to a small amount of impurity phase in the as-prepared sample.

The energy of magneto static interaction (dipole-dipole) depends on the magnetic moment is given by [44]

$$U_{max} = \frac{-\pi^2 D^3 M_S^2}{18} \qquad (5.10)$$

The energy of magneto static integration between the dipoles is -2.44×10^{-13}, -1.42×10^{-13}, and -1.57×10^{-13} for the MnS samples with 1:1, 1:2 and 2:1 molar ratio respectively. According to Neel, the magnetic moment of the particle relaxes to its equilibrium position with a relaxation time τ_N [45] given by the relation,

$$\tau_N = \tau_0 \exp\left(\frac{K_{eff}V}{k_B T}\right) \tag{5.11}$$

The estimated relaxation time τ_N for all the MnS sample is ~1ns.

5.4 Conclusions

The MnS nanocrystals have been successfully synthesized by using wet chemical synthesis route with 1:1, 1:2 and 2:1 precursor molar ratios. The XRD pattern reveals that the metastable β-MnS nanocrystals synthesized from 1:1 precursor molar ratio and metastable γ-MnS nanocrystals obtained from the sample with 1:2 and 2:1 precursor molar ratios. The average crystallite size using modified Scherrer equation, micro strain and dislocation density of as-synthesized MnS nanocrystals have been calculated from the XRD data. The average crystallite size is found to be 2.04 nm (1:1), 5.56 nm (1:2), and 1.74 nm (2:1). It has been observed that the dislocation density and microstrain decreases with increase in crystallite size. The rough surface spherical MnS particles are observed in the sample with 1:1 molar ratio and the diameter of the spheres varies within 1-3 μm. The sample prepared with 1:2 molar ratio shows the smooth spherical crystals, with an average diameter of about 3-5 μm. The agglomerates obtained in the sample with 2:1 molar ratio due to excess of Mn source. The FTIR spectral analysis reveals the characteristics peaks for Mn-S. UV-Visible absorbance spectra indicate that as-synthesized MnS nanocrystals are showing large blue shift compared to the bulk MnS. The optical band gap of MnS nanocrystals are found to be 4.72eV (1:1), 4.69eV (1:2) and 4.70eV (2:1).

The photoluminescence spectra reveals the two blue emission peaks around 409-420 nm and 433-442 nm with increased intensity as the crystallite size decreases. The blue emission peaks are assigned to the energy level of sulfur vacancies with the holes from the valance band and interstitial Mn from interstitial sulfur. The dielectric studies of the as-synthesized MnS nanocrystals reveal that both the dielectric constant and dielectric loss decrease with an increase in the frequency and increase with the increase in temperature. The impedance decreases rapidly with increase in frequency in the low frequency range and it reaches a constant value at high frequency range which is independent of frequency. The impedance spectra of the samples show the grain and grain boundary effects. It has been observed that the grain and grain boundary resistance increases with increase in sulfur molar concentration whereas decreases with increase in Mn molar concentration.

The bulk resistance decreases with increase in temperature, its behavior is likely to that of semiconductor. The modulus vs frequency plot clarifies the presence of both grain and grain boundary contribution to the conductivity of the material. In all temperatures, M' approaches nearly zero, confirming the presence of considerable electrode and ionic polarization in the temperature range. M' increases with frequency and reaches a constant value at higher frequencies. The peak frequency of M'' shifts towards the high frequency region as temperature increases, which implies conductivity relaxation suggesting that the dielectric relaxation, is activated thermally, in which a hopping process of charge carriers is predominant. The frequency of the applied field increases, hopping of charge carriers also increases thereby increasing the AC conductivity. The AC conductivity is found to be high for higher frequencies which confirm small polaron hopping in the samples. The raise in AC conductivity with respect to temperature is the characteristics for any semiconductor.

The value of activation energy decreases with the increase of frequency confirms that hopping conduction is the dominant current transport mechanisms. The measured magnetization curves shows that all MnS samples have clear paramagnetic behavior.

References

1. Knoll, B. and Keilmann, Nature 399 (1999) 134-137.
2. Shiladitya Sengupta, David Eavarone, Ishan Capila, Ganlin Zhao, Nicki Watson, Tanyel Kiziltepe, Ram Sasisekharan, Nature 436 (2005) 568-572.
3. L.P.S. Santos, E.R. Camargo, M.T. Fabbro, E. Longo, E.R. Leite, Ceram. Int. 33 (2007) 1205-1209.
4. Juan Beltran-Huarac, Javier Palomino, Oscar Resto, Jingzhou Wang, Wojciech M. Jadwisienczak, Brad R. Weiner and Gerardo Morella, RSC Adv. 4 (2014) 38103-38110.
5. Sheng Cao, Jinju Zheng, Jialong Zhao, Lin Wang, Fengmei Gao, Guodong Wei, Ruosheng Zeng, Linhai Tian and Weiyou Yang, J. Mater. Chem. C 1 (2013) 2540-2547.
6. C.D. Lokhande, A. Ennaoui, P.S. Patil, M. Giersig, M. Muller and K. Diesner, Thin Solid Films 330 (1998) 70-75.
7. S. S. Aplesnin, L. I. Ryabinkina, O. B. Romanova, D. A. Balaev, O. F. Demidenko, K. I. Yanushkevich, and N. S. Miroshnichenko, Physics of the Solid State 49 (2007) 2080-2085.
8. F.M. Michel, M.A.A. Schoonen, X.V. Zhang, S.T. Martin, J.B. Parise, Chem. Mater. 18 (2006) 1726-1736.
9. M. Dhanam, B. Kavitha, M. Shanmugapriya, Chalcogenide Lett. 6 (2009) 541-547.
10. A. Monshi, M.R. Foroughi, M.R. Monshi, World J. Nanosci. Eng. 2 (2012) 154-160
11. D.S. Rana, D.K. Chaturvedi, J.K. Quamara, J. Optoelectron. Adv. Mater. 11 (2009) 705-712.
12. N.Kandasamy, S.Saravanan, Deepak Ranjan Nayak, International Journal of Advanced Chemical Science and Applications 1 (2014) 25-30.
13. Pingtang Zhao, Qiumei Zeng, Xianliang He, Hao Tang, Kaixun Huang, Journal of Crystal Growth 310 (2008) 4268-4272.
14. Aniket Gade, Mahendra Rai, Sulabha Kulkarni, nternational Journal of Green Nanotechnology 3 (2011) 153-159.
15. Louse Barry, Mark Copley, Justin D. Holmes, David J. Otway, Olga Kazakova, Michael A. Morris, Journal of Solid State Chemistry 180 (2007) 3443-3449.
16. N. Moloto, M.J. Moloto, M. Kalenga, S. Govindraju, M. Airo, Optical Materials 36 (2013) 31-35.
17. O. Goede, W. Heimbrodt, M. Lamla, V. Weinhold, Phys. Status Solidi B 146 (1988) K65-K69.
18. L. Wang, S. Sivananthan, R. Sporken, R. Caudano, Phys. Rev. B 54 (1996) 2718-2722.
19. Yange Zhang, Zude Zhang, Shutao Wang, Xuchu Ma, Yitai Qian, Materials Chemistry and Physics 97 (2006) 365-370.
20. R. Viswanath, H.S. Bhojya Naik, G.S. Yashavanth Kumar, P.N. Prashanth Kumar, K.N. Harish, M.C. Prabhakara, Spectrochimica Acta Part A: Molecular and Biomolecular Spectroscopy 125 (2014) 222-227.
21. Xin Yu, Cao Li-yun, Huang Jian-feng, Liu Jia, Fei Jie, Yao Chun-yan, Journal of Alloys and Compounds 549 (2013) 1-5.
22. Berrard D. Cullity, S. R. Stock, Elements of X-Ray Diffraction, Prentice Hall, 3 edition, 2001.
23. J. Jiang, R. Yu, J. Zhu, R. Yi, G. Qiu, Y. He, X. Liu, Mater. Chem. Phys. 115 (2009) 502–506.

24. Remy Tappero, Pierre Wolfers, Albert Lichanot, Chemical Physics Letters 335 (2001) 449-457.
25. R V Waghmare, N G Belsare, F C Raghuwanshi, S N Shilaskar, Bull. Mater. Sci. 30 (2007) 167-172.
26. Mukul Pastor, P.K.Bajpai, R.N.P.Choudhary, Bull. Mat.Sc. 28 (2005) 199-203.
27. P.K.Bajpai, Kuldeep Ratre, Mukul Pastor, T.P.Sinha, Bull. Mat. Sc. 36 (2003) 461-464.
28. SKS Parashar, Swarat Chaudhuri, Satyendra Narayan Singh, Mahmood Ghoranneviss, Journal of Theoretical and Applied Physics 7:26 (2013).
29. S.K. Rout, Ali Hussian, J.S. Lee, I.W. Kim, S.I.Woo, Journal of Alloys and Compounds 477 (2009) 706-711.
30. Shaimaa A. Mohamed, A.A. Al-Ghamdi, G.D. Sharma, M.K. El Mansy, Journal of Advanced Research 5 (2014) 79-86.
31. Lei Zhang, Feng Liu, Kyle Brinkman, Kenneth L. Reifsnider, Anil V. Virkar, Journal of Power Sources 247 (2014) 947-960.
32. M.P. Dasari, K. Sambasiva Rao, P. Murali Krishna, G. Gopala Krishna, Acta Physica Polonica A 119 (2011) 387-394.
33. A Molak, M Paluch, S Pawlus, J Klimontko, Z Ujma, I Gruszka, Journal of Physics D: Applied Physics 38 (2005) 1450-1460.
34. Ghada E. El-Falaky, Osiris W. Guirguis, Nadia S. Abd El-Aal, Progress in Natural Science: Materials International 22 (2012) 86-93.
35. Alo Dutta, T.P. Sinha, P. Jena, S. Adak, Journal of Non-Crystalline Solids 354 (2008) 3952-3957.
36. Rajalingam Venkatesan, Subramaniam Velumani, Mohamed Tabellout, Nicolas Errien, Abdelhadi Kassiba, Journal of Physics and Chemistry of Solids 74 (2013) 1695-1702.
37. S.K. Rout, Ali Hussian, J.S. Lee, I.W. Kim, S.I. Woo, Journal of Alloys and Compounds 477 (2009) 706-711.
38. M. S. Hossain, R. Islam, K. A. Khan, Chalcogenide Letters 5 (2008) 1-9.
39. Adem Tataroglu, Gazi University Journal of Science 26 (2013) 501-508.
40. D.J. Sellmyer, Ralph Skomski, Advanced Magnetic Nanostructures, Springer Science & Business Media, 2006.
41. T. Kurz, L. Chen, F. J. Brieler, P. J. Klar, H.A.Krug von Nidda, M. Fröba, W. Heimbrodt, A. Loidl, Phys. Rev. B 78 (2008) 132408.
42. Jian-feng WEN, Ye ZHUANG, Nu-jiang TANG, Li-ya LU, Wei ZHONG, You-wei DU, New Carbon Materials 28 (2013) 66-70.
43. C. Gumu, A. Bayri, C. Uluta, M. Karakaplan, Y. Ufuktepe, Journal of Optoelectronics and Advanced Materials 8 (2006) 261-265.
44. H.R. Bertorello, P.G. Bercoff, M.I. Oliva, Journal of Magnetism and Magnetic Materials 269 (2004) 122-130.
45. Yu. P. Kalmykov, W. T. Coffey, S. V. Titov, Physics of the Solid State 47 (2005) 272-280.

Chapter VI

INFLUENCE OF Cu-DOPING ON STRUCTURAL, OPTICAL, ELECTRICAL AND MAGNETIC PROPERTIES OF CHEMICALLY SYNTHESIZED MnS NANOCRYSTALS

6.1 Introduction

The manganese chalcogenides are diluted magnetic semiconductors. It has outstanding magneto-optical properties [1]. The MnS exhibits anti-ferromagnetic behavior with a transition temperature T_N=130K due to the correlations between the Mn^{2+} spins, and a paramagnetic moment p_{eff} =5.6 μB [2]. In the paramagnetic phase, the MnS is a p-type semiconductor with an activation energy E_a = 0.3 eV [3]. The metastable γ-MnS and Cu-doped MnS nanocrystals have been synthesized by wet chemical synthesis technique. The effect of Cu-doping on the structural, optical, electrical and magnetic properties has been investigated. The synthesized MnS and Cu-doped MnS nanocrystals having reduced dimensions allow the possibility of controlled quantum confinement. Owing to their attractive properties, these materials can be used in anode material for Li-ion batteries [4], optoelectronic devices, buffer material in solar cell and magneto-optical devices [5].

The Cu-doped MnS with high electrical conductivity and superparamagnetism extends their application in the blue green light emitters [6], sensors, and electro-magnetic resonance applications [7].

6.2 Experimental Details

6.2.1 Materials

The chemicals used for the synthesis of MnS samples were of high purity procured from Merck supplier. Therefore no further purification was considered necessary and hence used as received. The chemicals used are manganese acetate [$Mn(CH_3COO)_2$], copper acetate monohydrate [$Cu(CH_3COO)_2H_2O$], thioacetamide [CH_3CSNH_2], ammonium chloride [NH_4Cl], triethanolamine [$N(CH_2CH_2OH)_3$] and trisodium citrate [$C_6H_5Na_3O_7$].

6.2.2 Synthesis

The Fig. 6.1 shows various steps of experimental procedures for synthesizing undoped and Cu-doped MnS nanocrystals. The MnS nanocrystals were synthesized as follows: 20ml of 1M manganese acetate aqueous solution was mixed with 2mL of triethanolamine and 20mL of 1M ammonium chloride under vigorous stirring. Then, 0.4mL of 0.7M trisodium citrate and 20mL of 1M thioacetamide were slowly added one by one in the mixture solution. The solution became clear and homogeneous. The pH of the solution was adjusted to 9.5 and stirred for 2h. The mixture was put into a 100ml two necked round bottom flask and refluxed at 65°C for 3h. The final solution was centrifuged, washed and oven dried at 60°C for 8h. The Cu-doped MnS nanocrystals were synthesized as follows: 20ml of 0.01M copper (II) acetate monohydrate was added drop wise in 20ml of 1M manganese acetate solution under vigorous stirring. Then, 2mL of triethanolamine and 20mL of 1M ammonium chloride were added one by one.

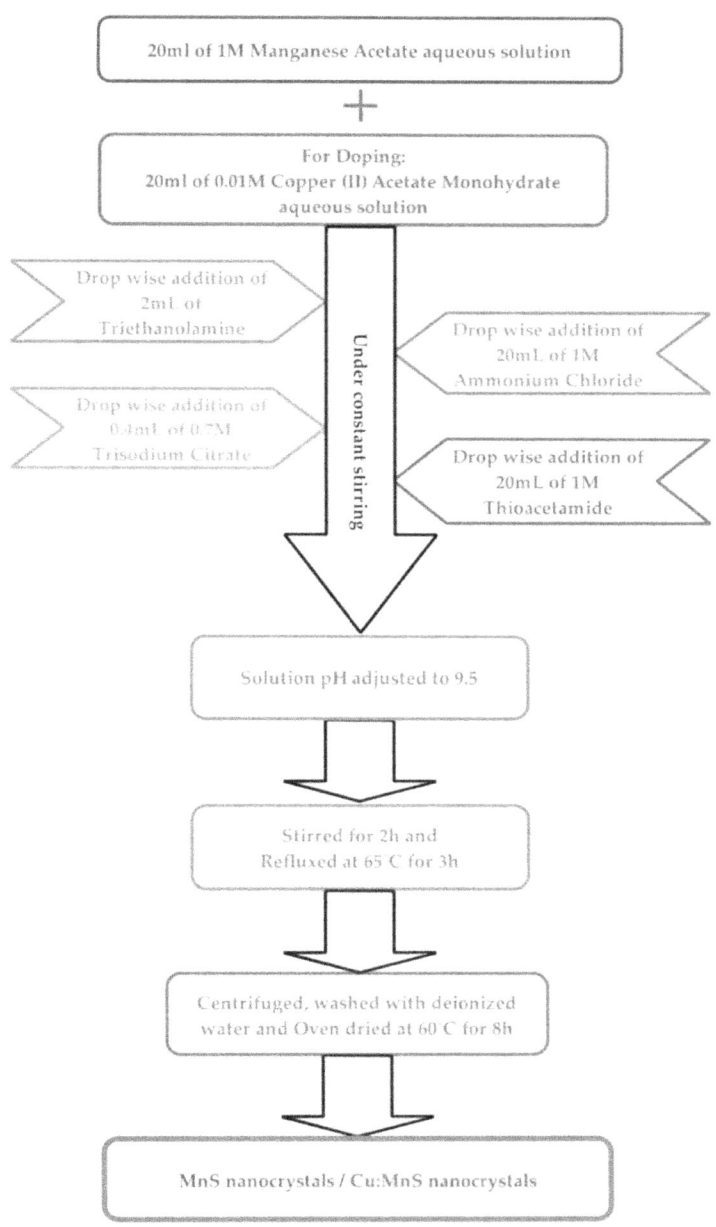

Fig. 6.1: Various steps involved in the experimental procedures for undoped and Cu-doped MnS nanocrystals

Under stirring, 0.4mL of 0.7M trisodium citrate and 20mL of 1M thioacetamide solutions were slowly dropped into the above mixture. The final solution at pH~9.5 was magnetically stirred well for 2h and refluxed at 65°C for 3h. The resulting precipitates were filtered and oven dried at 60°C for 8h.

6.3 Results and Discussion

6.3.1 Structural Properties

The structural properties of MnS and Cu-doped MnS nanocrystals have been carried out by the powder XRD analysis and transmission electron microscopy (TEM) analysis. The FTIR spectrum is used to study the stretching and bending frequencies of molecular functional groups in the sample.

6.3.1.1 XRD Analysis

The Fig. 6.2(a) shows the XRD pattern of pure MnS nanocrystals, the peaks at 2θ of 26.0°, 27.8°, 29.6° and 45.6° corresponding to the (100), (002), (101), and (110) diffraction planes. The lattice constants are found to be a = 3.95A° and c = 6.41A° [8-11] and are in agreement with the standard diffraction data of γ-MnS (JCPDS: 40-1289) [12].

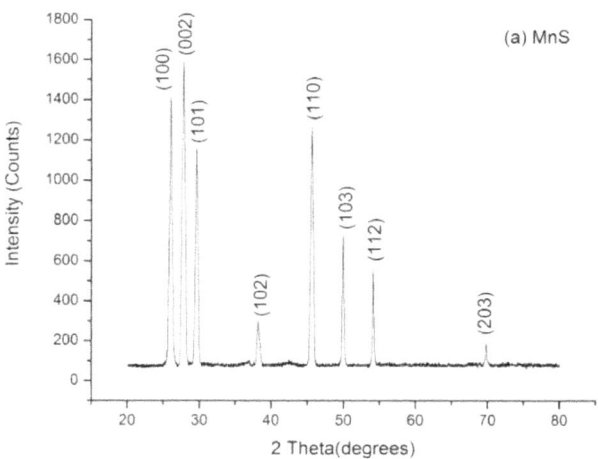

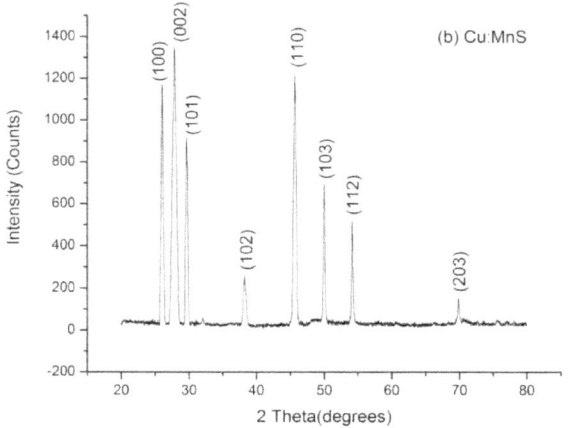

Fig. 6.2: XRD patterns of (a) undoped MnS and (b) Cu-doped MnS nanocrystals

The diffraction pattern of Cu-doped MnS sample is shown in Fig. 6.2(b). It is observed that the diffraction peaks of Cu-doped MnS is shifted to the higher angle of 2θ by 0.06° compared to the undoped MnS. The intensity of the diffraction peaks are decreased and peak at (002) is broadened as a result of Cu-doping. The calculated lattice constants are a = 3.93A° and c = 6.36A°. The decreasing trend of lattice constant is due to the substitution of Cu atoms for Mn atoms. The average crystallite size (D) and strain broadening (ε) are calculated using Williamson-Hall equation [13].

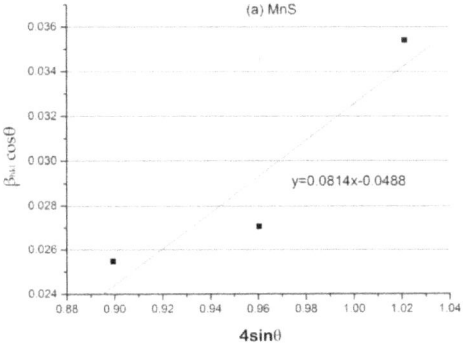

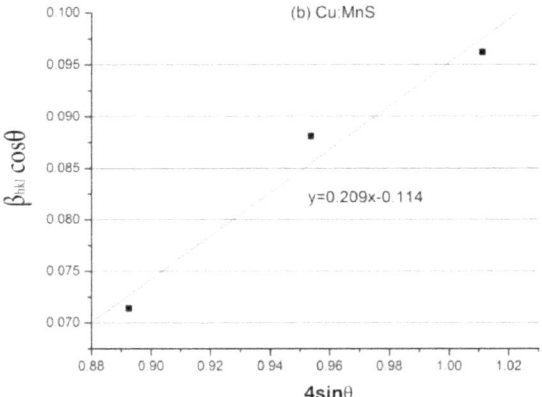

Fig. 6.3: W-H plots for (a) undoped MnS and (b) Cu-doped MnS nanocrystals

$$\beta_{hkl} \cos\theta = \frac{K\lambda}{D} + 4\varepsilon \sin\theta \qquad (6.1)$$

where, β_{hkl} is instrument-corrected broadening corresponding to the Bragg peaks, θ is Bragg's angle, K is shape factor (0.9), λ is wavelength of Cu$k\alpha$ radiation, D is crystalline size and ε is strain broadening.

The W-H plots are drawn between $\beta_{hkl} \cos\theta$ (in y-axis) and $4\sin\theta$ (in x-axis) for the wurtzite phase peaks is shown in Fig. 6.3(a & b). The linear fit on the plot gives the crystallite size and strain from the y-intersect and slope respectively [14]. The average crystallite size and strain of the undoped and the Cu-doped MnS samples are calculated as 2.88 nm & 0.081 and 1.27 nm & 0.209 respectively. It is clear that the lattice strain increases with decreasing crystallite size. The dislocation density (δ) is calculated by using the expression [15]:

$$\delta = 1/D^2 \qquad (6.2)$$

The dislocation density of MnS and Cu-doped MnS nanocrystals is found to be 0.1198 nm^{-2} and 0.6759 nm^{-2} respectively. The specific surface area is increased as the particle size becomes small. The specific surface area is important for the industrial process and chemical reaction.

Even with the same material that has the same weight and volume, the surface activity and adsorption volume are changed according to the specific surface area. The specific surface area of the MnS samples is calculated using formula as

$$SSA = \frac{6*10^3}{D.\rho} \quad (6.3)$$

where, D is the crystallite size and ρ is the density of MnS (3990 kgm^{-3}). The specific surface area of undoped and Cu-doped MnS samples is calculated as 387.56 m^{2}g^{-1} and 1184.06 m^{2}g^{-1} respectively.

6.3.1.2 TEM Analysis

The particle size, size distribution and morphology of the as-prepared MnS powders are analyzed by transmission electron microscopy. TEM images can also be used to judge whether good dispersion is achieved or agglomeration is present in the system. The TEM micrographs of the undoped MnS nanocrystals are shown in Fig. 6.4(a). The TEM images confirm the particles are spherical in shape and obviously well dispersed.

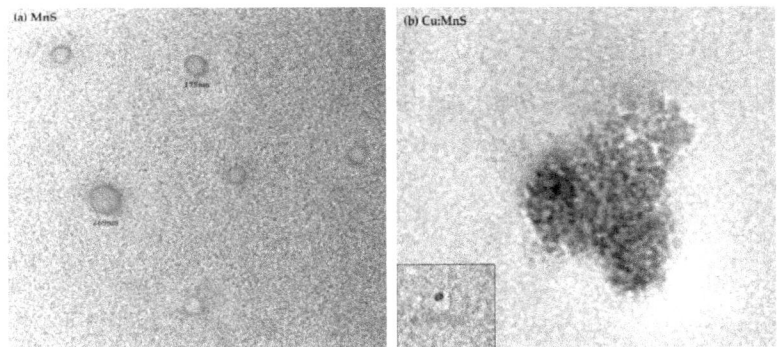

Fig. 6.4: TEM images of (a) undoped MnS and (b) Cu-doped MnS nanocrystals

The estimated undoped MnS sample shows size distribution approximately 175 nm to 300 nm, which is in agreement with XRD results.

Fig. 6.4 (b) exhibits the size of the particles on Cu-doping is found to be decreased significantly may be due to the small ionic radii of Cu^{2+} (0.057 nm) when compared to Mn^{2+} ion (0.080 nm). The particle size of Cu-doped sample obtained from TEM image is around 15 nm.

6.3.1.3 FTIR Spectra Analysis

The formation of nanocrystalline MnS and Cu-doped MnS structures are further confirmed by FTIR spectral analysis as shown in Fig. 6.5(a & b). The FTIR spectra for MnS and Cu-doped MnS vibration peaks can be observed in the range of 4000-400 cm^{-1}. The broad absorption peak in the range of 3400-3900 cm^{-1} corresponding to O-H group which indicates the existence of water absorbed in the surface of nanocrystals [16]. The bands at 3308 cm^{-1} and 3452 cm^{-1} are due to the N-H stretching. The band observed at 2966 cm^{-1} and 2974 cm^{-1} indicates an asymmetry in stretching mode of CH_2 group [17]. The C-C vibration occurs at about 2314 cm^{-1}. The peaks at 1562 cm^{-1} and 1548 cm^{-1} are O-H bending vibrations [16]. The absorption band appeared at 1398 cm^{-1} and 1392 cm^{-1} corresponds to CH_2 bending. The peaks at 1253 and 1219 cm^{-1} correspond to N–H bending.

The appearing broadband peaks at 590 cm^{-1} is attributed to Mn-S-Cu. The bands are observed at 880 cm^{-1} and 879 cm^{-1} indicates the vibrational modes of sulfide ions in the crystal. An absorption band near 560 cm^{-1} is associated with vibrations of MnS structural units [18]. The absorption peaks found to be shifted slightly towards the higher frequency side which indicates the introduction of dopant in MnS nanocrystals.

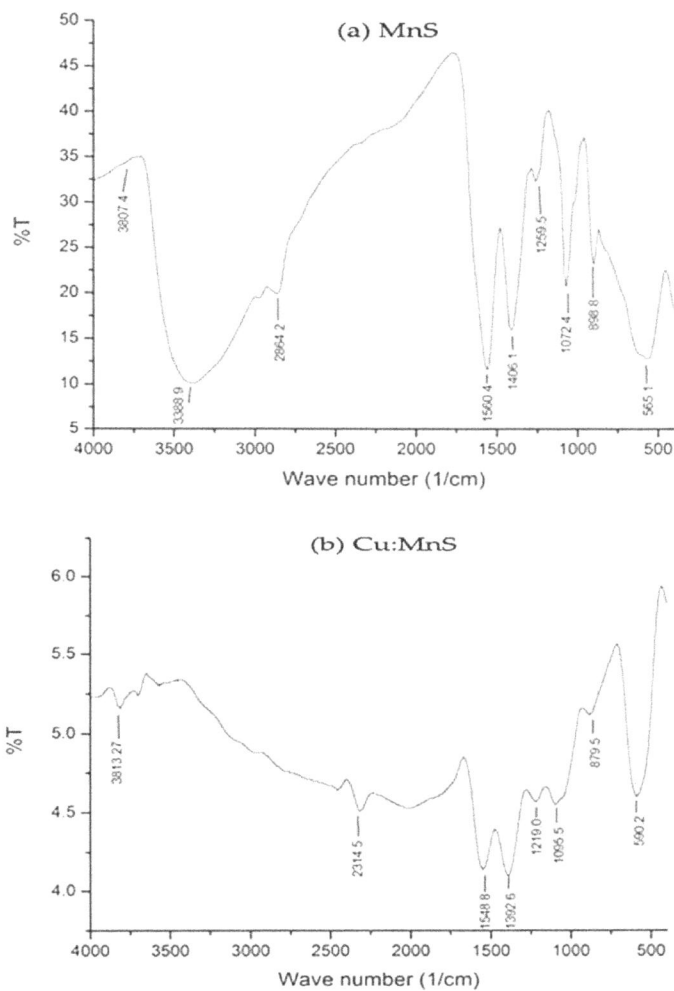

Fig. 6.5: FTIR spectra of (a) undoped MnS and (b) Cu-doped MnS nanocrystals

6.3.2 Optical Properties

The optical properties of the samples are investigated by measuring the UV-Visible absorbance and photoluminescence spectra at room temperature. The influences of Cu-doping on the optical properties of manganese sulfide nanocrystals are studied.

6.3.2.1 UV-Visible Absorption Spectra Analysis

Fig. 6.6(a) shows the optical absorption spectra of MnS nanocrystals and Cu-doped MnS nanocrystals. It is found that the absorption spectra of the Cu-doped MnS nanocrystals exhibit a blue shift when compared to that of the MnS nanocrystals. This may be attributed to reduction in particle size due to the incorporation of Cu^{2+} ions on a Mn^{2+} lattice site [19]. The large blue shifted peaks are concerned with a quantum size effect which is due to the confinement of electrons and holes in crystal lattice [20, 21].

The optical band gap of these materials is determined from optical absorption spectra using the relation [22, 23],

$$E_g = hc/\lambda \ (eV) \qquad (6.4)$$

where, c is speed of light, h is the Planck's constant, λ is the cut off wavelength and E_g is the optical band gap of the material.

The optical band gap is calculated and is found to be 4.30 eV and 4.74 eV for MnS and Cu-doped MnS nanocrystals respectively. It reveals that the band gap value is raised by the Burstein-Moss effect due to introduction of Cu-dopant. It implies an increase in electrical conductivity due to increase in carrier concentration in the sample [24].

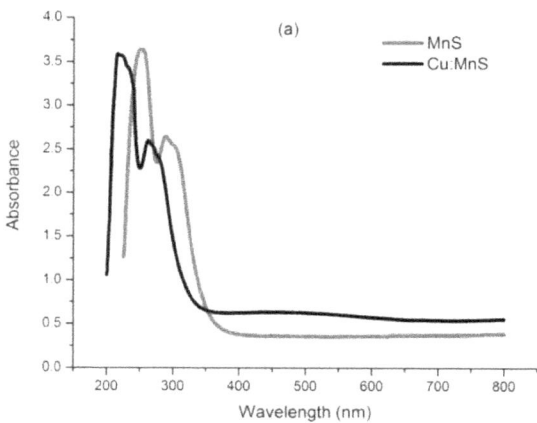

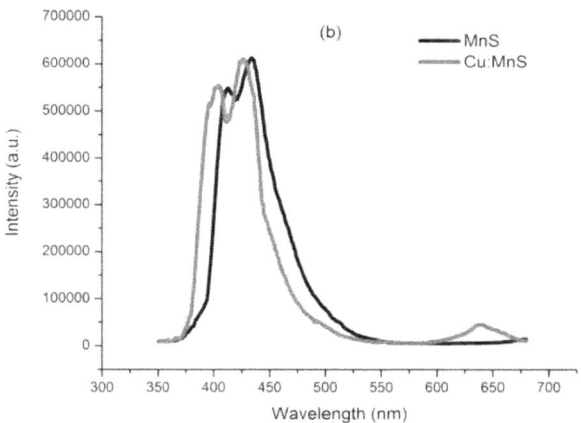

Fig. 6.6: Optical properties of MnS and Cu-doped MnS nanocrystals (a) UV-Visible absorbance spectra (b) Photoluminescence (PL) spectra

6.3.2.2 PL Spectra Analysis

The Photoluminescence spectra (PL) of MnS and Cu-doped MnS nanocrystals are recorded at RT with excitation at 350 nm as shown in Fig. 6.6 (b). The PL spectra of the samples exhibit two peaks at 419 nm and 441 nm, 413 nm and 434 nm are attributed to the blue emissions of undoped and cu-doped MnS nanocrystals respectively. PL intensity slightly decreases with Cu-doping because of decreasing grain size and decreasing crystallinity of the sample. The emission peaks at 419 nm and 413 nm can be attributed to the recombination of electrons at the sulfur vacancy related donor and valance band. The observed PL peaks at 441 nm and 434 nm are assigned to the recombination of charge carriers in deep traps of surface localized stated and a photo-generated hole caused by surface defects [25]. When Cu^{2+} ions are doped into MnS nanocrystals, more defect states are introduced. Therefore, new peak appeared in the longer wavelength side. The emission bands centered at 594 nm reveals the substitution of Cu^{2+} ions into the MnS host lattice.

By comparing the two spectra, the Cu-doped MnS spectrum is red shifted from the undoped MnS.

6.3.3 Electrical Properties

The electrical properties of synthesized nanocrystals are studied using impedance spectroscopy. The effects of Cu-doping on dielectric, impedance, electric modulus and AC conductivity phenomena of the material with respect to temperature and frequency are analyzed.

6.3.3.1 Dielectric Studies

The frequency dependence of the dielectric constant at different temperatures is observed in Fig. 6.7(a & b). The dielectric constant is complex in AC field and is given by

$$\varepsilon = \varepsilon' - j\varepsilon'' \qquad (6.5)$$

where ε' represents the stored energy and ε'' represents the dissipated energy. The dielectric constant decreases with increase in frequency. This decrease is rapid at lower frequencies and slower at higher frequencies. It indicates that the dispersion due to Maxwell-Wagner type interfacial polarization which agrees with Koop's phenomenological theory [26]. Based on this model, the dielectric materials are composed of conducting grains separated by poorly conducting grain boundaries.

If the frequency of field increases, the probability of electrons reaching the grain boundaries decreases and hence dielectric constant decreases due to decreases in polarization. It indicates the dielectric relaxation which is related to MnS-based polarization [27]. The value of dielectric constant increases with Cu-doping due to reduction of grain size. Reducing grain size has the effect of increasing domain wall concentration which increases the permittivity. The increase in dielectric constant with increasing temperature can be attributed to the thermal activation of electron exchange as a result the space charge polarization occurs near the grain boundary [28].

Loss tangent or loss factor tanδ represents the energy dissipation in the dielectric system. The variations of dielectric loss as a function of frequency are shown in Fig. 6.8(a & b). As the frequency increases the dipoles are less able to rotate and maintain phase with the applied field, they reduce their contribution to the polarization field. Therefore, the dielectric loss is observed to decrease with increasing frequency. The drop in resistivity due to Cu-doping [29] which would have caused further rise in dielectric constant. Contributions to the overall dielectric constant and dielectric loss from different polarization mechanisms can be related to the composition, porosity, grain morphology, frequency and temperature of the dielectric and their relative effect can be held responsible for the observed dielectric behaviour of the material.

6.3.3.2 Impedance Studies

The total impedance Z^* of the equivalent circuit is defined by the relation

$$Z^* = Z' - jZ'' \tag{6.6}$$

where Z' is the real part of impedance which is related to a pure resistance R, Z'' is the imaginary part of impedance that can be related to a capacitance C where $Z'' = 1/j\omega C$ [30].

In Fig. 6.9(a & b), the impedance values are typically higher in low frequency region and decreases gradually with increasing frequency for all temperature. At low frequency, the magnitude of Z decreases with increasing temperature indicating that the NTCR-type behavior of semiconductors [31]. The value of $|Z|$ appears to merge in the high frequency region for both the samples with increasing temperature which reduces the barrier properties of the material. The total impedance of Cu-doped MnS is lower than the MnS which indicates doped sample having increased conductivity.

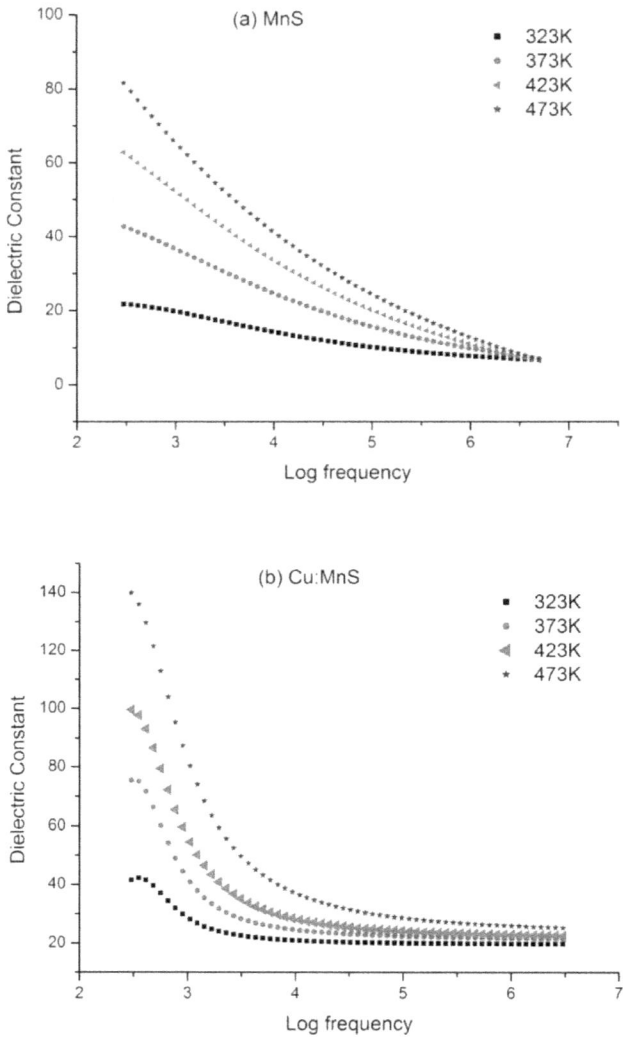

Fig. 6.7: Variation of dielectric constant with frequency at different temperatures for (a) undoped MnS and (b) Cu-doped MnS nanocrystals

The impedance data of material having resistive and capacitive components is plotted in the complex plane. It appears in the form of a sequence of semicircles representing electrical phenomenon due to grain, grain boundary and interfacial phenomenon of the material.

The grains are effective in high frequency region while the grain boundaries are effective in low frequency region.

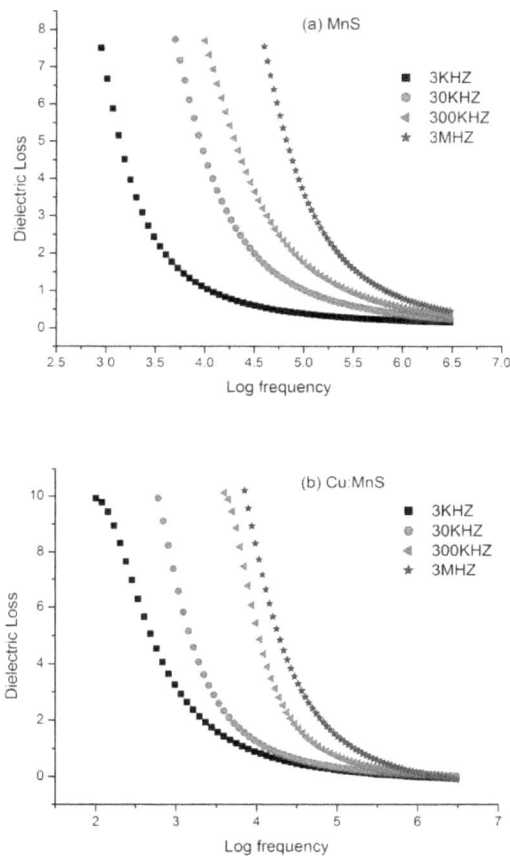

Fig. 6.8: Variation of dielectric loss with frequency at different temperatures for (a) undoped MnS and (b) Cu-doped MnS nanocrystals

The Fig. 6.10 (a & b) shows complex impedance plot (Cole-Cole plot) of Z' verses Z" are useful for determining the dominant resistance of the sample, where real Z' represents the resistive part and the Z'' values represents the capacitive part. The impedance spectra of the undoped and Cu-doped MnS samples are apparently single depressed half circles.

The equivalent circuit may be considered as two parallel RC elements connected in series and giving rise to two arcs in complex plane: one for the grain and another for the grain boundary. A small semicircle at high frequencies indicates the bulk (grain) properties of the materials arising due to a parallel combination of bulk resistance and bulk capacitance. The large semicircle at low frequencies indicates the presence of grain boundary arising due to a parallel combination of grain boundary resistance and capacitance. The intercepts of the semicircular arc on real axis gives the grain (R_g) and grain boundary (R_{gb}) resistance of the material [32]. The capacitance of the material can be calculated by using,

$$\omega_{max}\tau = 2\pi f_{max} RC = 1 \tag{6.7}$$

where ω_{max} is the maximum angular frequency of semicircle and τ is the relaxation time. The values of equivalent circuit parameters calculated from the impedance plane are given in Table 6.1.

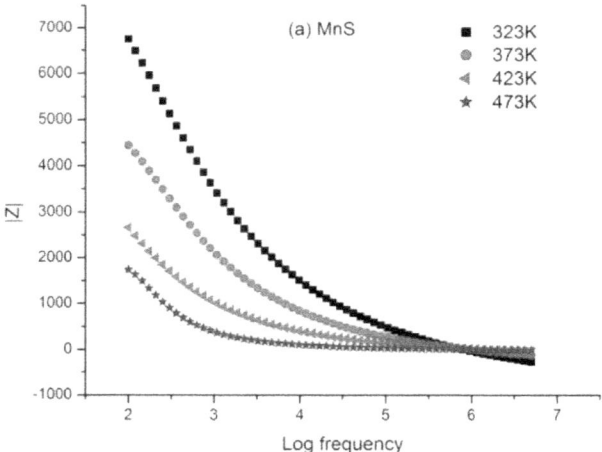

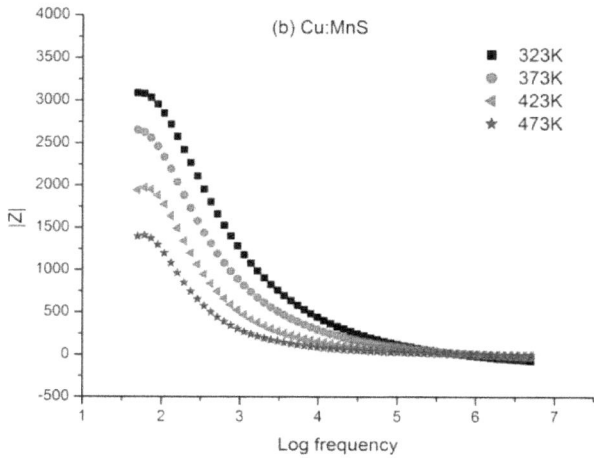

Fig. 6.9: Variation of impedance |Z| with frequency at different temperatures for (a) undoped MnS and (b) Cu-doped MnS nanocrystals

Table 6.1: Values of the equivalent circuit parameters deduced from the impedance spectra for the undoped MnS and Cu-doped MnS nanocrystals

Samples and Factors		Grain resistance (R_g) Ω	Grain boundary resistance (R_{gb}) Ω	Grain capacitance (C_g) F	Grain boundary capacitance (C_{gb}) F
MnS	323K	291	3758	4.88E-07	3.78E-08
	373K	282	2226	8.22E-07	1.04E-07
	423K	267	1239	1.86E-06	4.02E-07
	473K	253	470	5.77E-06	3.11E-06
Cu:MnS	323K	64	2430	4.12E-06	1.08E-07
	373K	56	1711	7.79E-06	2.55E-07
	423K	49	1223	1.83E-05	7.31E-07
	473K	36	462	5.20E-05	4.05E-06

The relaxation time decreases with the increase in temperature is shown in Fig. 6.11. It is indicating that the relaxation phenomenon is thermally activated. The dielectric relaxation time is closely related to the electrical conductivity.

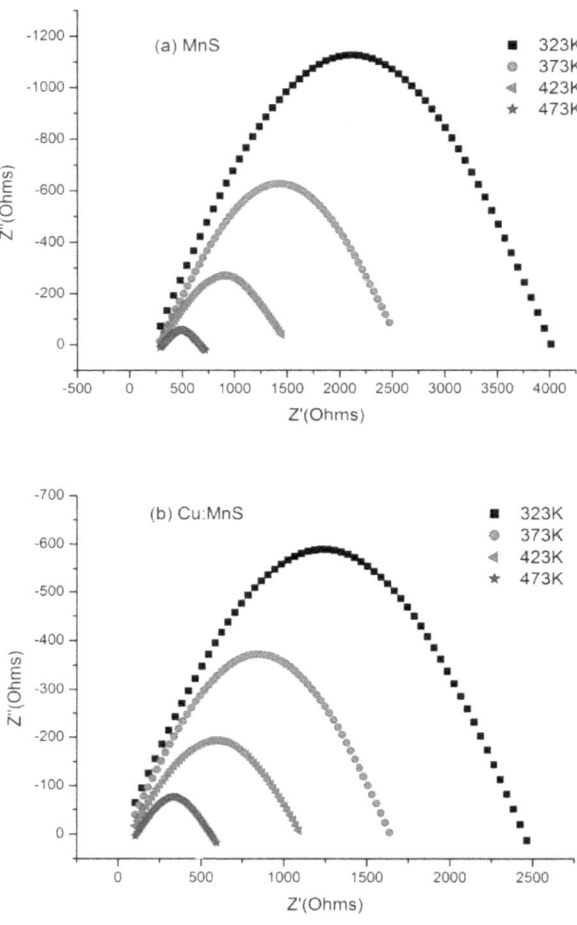

Fig. 6.10: Complex impedance plots for (a) undoped MnS and (b) Cu-doped MnS nanocrystals

6.3.3.3 Electric Modulus Studies

The electric modulus is the reciprocal of the permittivity $M^* = 1/\xi^*$. Although it was originally introduced by Macedo [33] to study space charge relaxation phenomena, M^* representation is now widely used to analyze ionic conductivities [34].

The electric modulus corresponds to the relaxation of an electric field in the material when the electric displacement remains constant, thus the electric modulus represents the real dielectric relaxation process, which can be defined by the equation,

$$M^* = M' + jM'' \quad (6.8)$$

where M' is the real and M" the imaginary electric modulus.

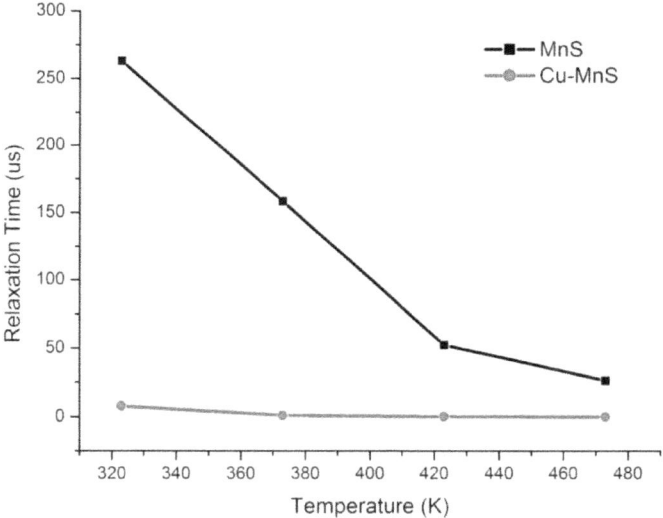

Fig. 6.11: Relaxation time with temperature plot for the undoped MnS and Cu-doped MnS nanocrystals

Fig. 6.12(a & b) shows the variation of M' as a function of frequency for the MnS and Cu-doped MnS NCs at selected temperatures. M' approaches to zero in the low frequency region, and a continuous dispersion on increasing frequency may be contributed to the conduction phenomena due to short range mobility of charge carriers. This implies the lack of a restoring force for flow of charge under the influence of a steady electric field [35]. This confirms elimination of electrode effect in the material.

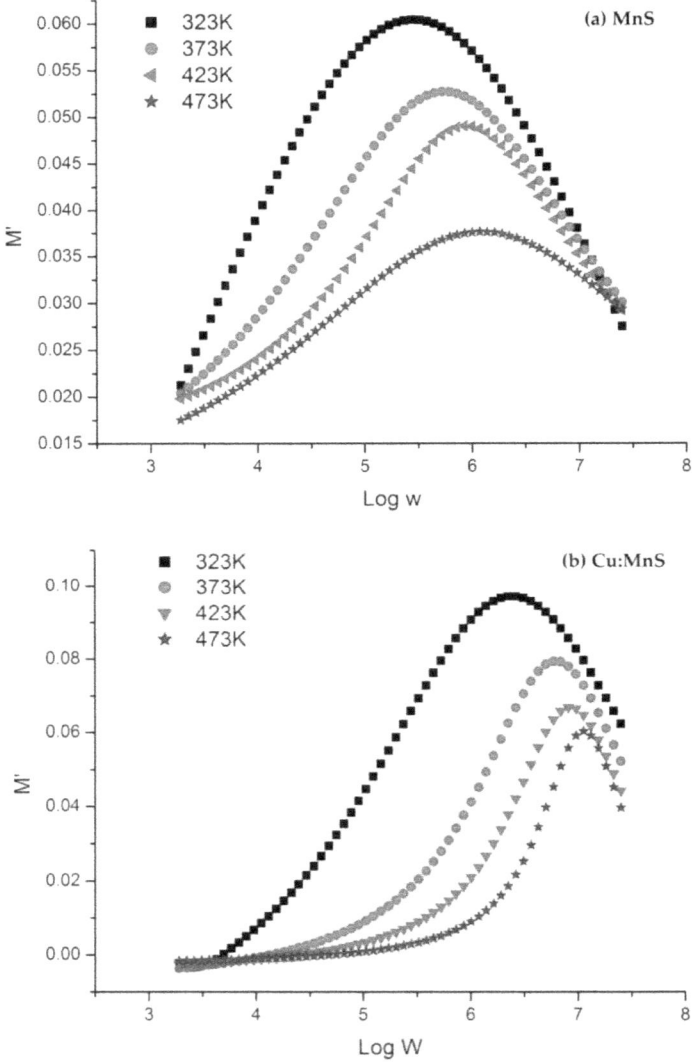

Fig. 6.12: Frequency dependence of M' at various temperatures for (a) undoped MnS and (b) Cu-doped MnS nanocrystals

Fig. 6.13(a & b) shows the variation of imaginary part of electric modulus (M'') with frequency for MnS and Cu-doped MnS NCs at selected temperatures.

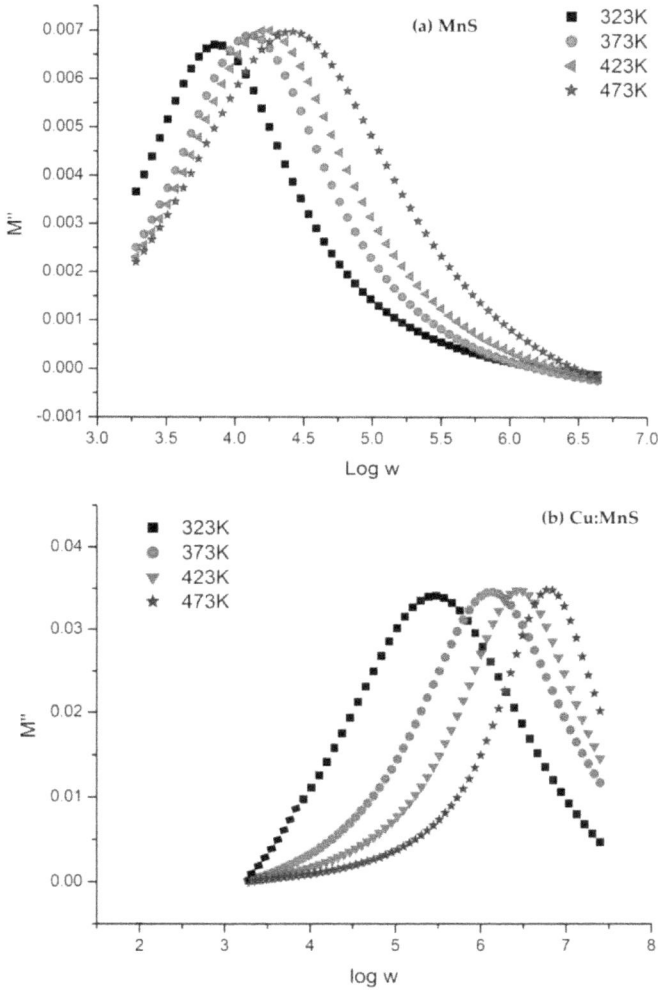

Fig. 6.13: Frequency dependence of M" at various temperatures for (a) undoped MnS and (b) Cu-doped MnS nanocrystals

The maxima M"$_{max}$ shifts towards higher frequencies side with rise in temperature as well as Cu-doping ascribing correlation between motions of mobile ions [36] and suggests that the dielectric relaxation is thermally activated process.

The asymmetric peak broadening indicates the spread of relaxation times with different time constant, and hence relaxation is of non-Debye type. The low frequency peaks show that the ions can move over long distances whereas high-frequency peaks merge to spatially confinement of ions in their potential well. The nature of modulus spectrum suggests the existence of hopping mechanism of electrical conduction in the materials.

6.3.3.4 AC Conductivity Studies

AC conductivity analysis measures the electric properties of a material as a function of frequency and temperature. The AC conductivity can be calculated using the relation [37],

$$\sigma_{ac} = \omega \varepsilon_0 \varepsilon_r \tan \delta \qquad (6.9)$$

where, $\omega=2\pi f$, f is the frequency, ε_0 is the permittivity of free space, ε_r is the relative dielectric constant and $\tan\delta$ is the dissipation factor.

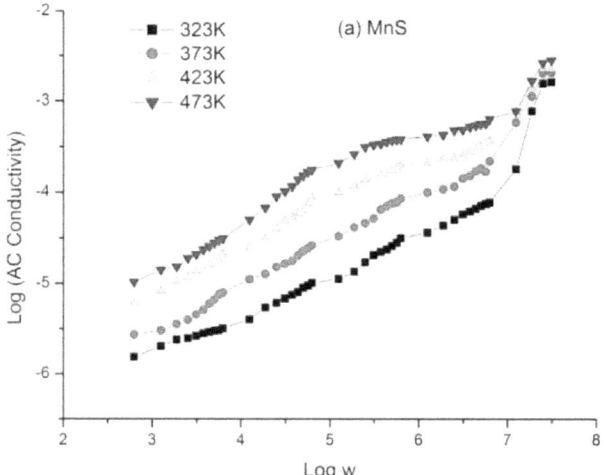

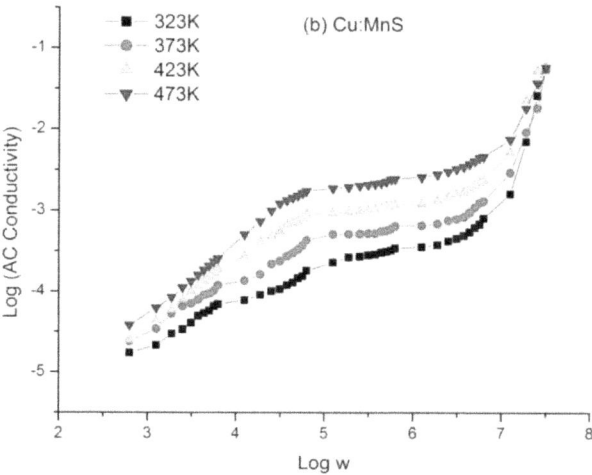

Fig. 6.14: Variation of AC conductivity with frequency at different temperatures for (a) undoped MnS and (b) Cu-doped MnS nanocrystals

The angular frequency dependence of AC conductivity at different temperatures is shown in Fig. 6.14(a & b). It is observed that AC conductivity gradually increases with the increase in frequency of applied field because the increase in frequency enhances the electron hopping frequency of charge carriers. A similar behavior is observed in all other temperatures. The number of charge carriers might be less in low frequencies. The number of charge carriers with low barrier height is more in high frequency region [38]. At low temperatures, the electron hopping conduction dominates the conductivity and ionic hopping makes little contribution. Further increasing the temperature, conductivity shows increasing behavior that indicates NTCR effect like the semiconductor. It is related to the thermally activated polaron hopping in the samples. It is observed that AC conductivity increases with the Cu-doping to MnS. Thus, the dopant increases the sulfur vacancies which results in an increase of free electron density and conductivity.

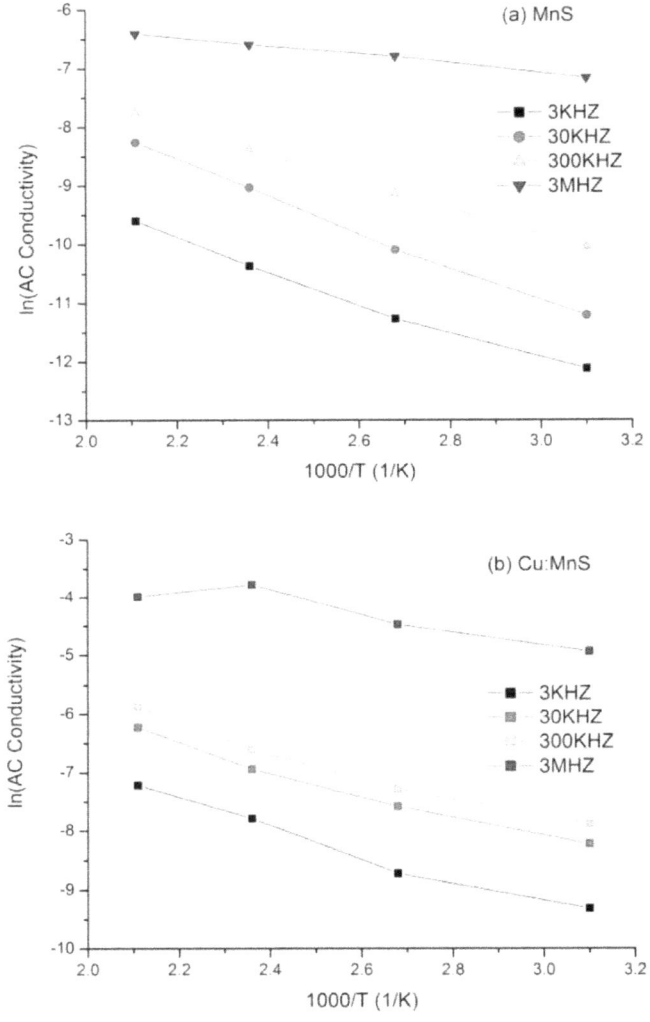

Fig. 6.15: Inverse temperature dependence of AC conductivity at various frequencies for (a) undoped MnS and (b) Cu-doped MnS nanocrystals

The inverse temperature dependence of the AC conductivity at various frequencies is observed in Fig. 6.15(a & b) from which the activation energy of samples is obtained. The temperature dependence of the conductivity is described by using the Arrhenius equation [39],

$$\sigma = \sigma_0 \exp(-E_a / k_B T) \qquad (6.10)$$

where, σ_o is the high temperature limit of conductivity, E_a is activation energy, k_B is Boltzman constant and T is temperature. The activation energies of MnS and Cu-doped MnS nanocrystals obtained at 373K are found to be 0.25 eV and 0.17 eV respectively. The activation energy decreases with increasing temperature due to the minimum energy required to overcome potential barrier. In Cu-doped MnS, the more dopant atoms occupy lattice sites of manganese atoms resulting in more charge carriers.

6.3.4 Magnetic Properties

The Fig. 6.16 exhibits the paramagnetic behavior for MnS nanocrystals and the superparamagnetic behavior for Cu-doped MnS nanocrystals at room temperature. The saturation magnetization, remanent magnetization, and intrinsic coercive force for the both undoped and Cu-doped MnS samples are listed in Table 6.2.

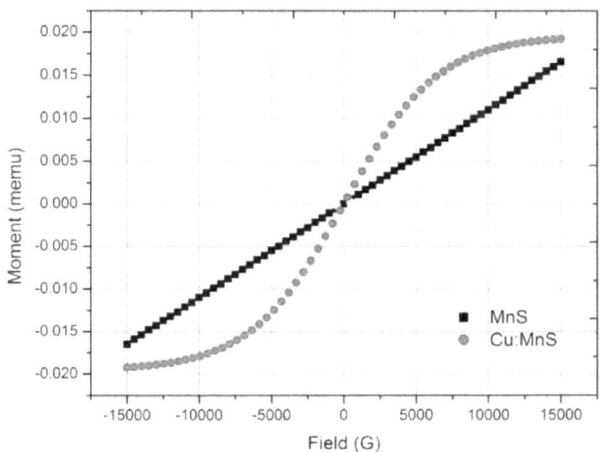

Fig. 6.16: Magnetization curve as a function of applied magnetic field at RT for the undoped MnS and Cu-doped MnS nanocrystals

The saturation magnetization increases and intrinsic coercivity decreases in Cu-doped MnS sample. The particle size is reduced below the single domain limit which exhibits super paramagnetism at room temperature [40]. The ratio of remanence to saturation magnetization (Mr/Ms) is 0.0033 and 0.289 for undoped and Cu-doped MnS respectively. For the paramagnetic material the value of Mr/Ms is less than 0.25 [41].

Table 6.2: M_s, M_r, M_c and Magnetic moment of undoped MnS and Cu-doped MnS nanocrystals

Samples	M_s (emu/g)	M_r (memu/g)	M_c (Oe)	Effective magnetic moment (μB)
MnS	0.256	0.848	7.3	1.42
Cu:MnS	14.8	4.29	1.46	1.55

The effective magnetic moment is estimated from [42],

$$\mu_{eff} = \sqrt{3kT\chi_m / N_A} \qquad (6.11)$$

The effective magnetic moment is found to be 1.42 µB and 1.55 µB for undoped and Cu-doped MnS respectively.

The magnetic moment is less than the spin only value (5.91 µB) [43] because of weak coupling of the unpaired electrons on the copper atoms. The estimation of K_{eff} and particle volume from the following relations,

$$K_{eff} = \frac{H_c M_s}{2} \qquad (6.12)$$

$$V = \frac{25 K_B T_B}{K_A} \qquad (6.13)$$

where, K_{eff} and K_A are effective magnetic anisotropy constant [44], H_C is coercivity, M_S is saturation magnetization, K_B is Boltzmann constant (1.38×10^{-23} J/K), T_B is blocking temperature in K, and V is particle volume.

According to Neel, the magnetic moment of the particle relaxes to its equilibrium position with a relaxation time (τ_N) given by the relation [45],

$$\tau_N = \tau_0 \exp\left(\frac{K_{eff}V}{k_B T}\right) \qquad (6.14)$$

The estimated relaxation time τ_N for the both sample is ~1ns.

6.4 Conclusions

Wet chemical synthesis route has been successfully used to synthesize manganese sulfide and copper-doped manganese sulfide nanocrystals. The XRD patterns show that the prepared samples have been γ-wurtzite phase nanocrystals. The average crystallite size is 2.88 nm and 1.27 nm for the undoped and Cu-doped MnS nanocrystals calculated by Williamson-Hall method. The optical energy band gap is found to be 4.30eV and 4.74eV for undoped and Cu-doped MnS nanocrystals respectively. The room temperature photoluminescence spectra of as-prepared samples reveal the broad blue emission peaks. FTIR analysis confirms the formation of MnS and Cu-doped MnS samples.

In the low frequency region, the dielectric constant and dielectric loss decreases with the increase in frequency, whereas in the high frequency region, it shows the frequency independent behavior and at high frequency dielectric loss is constant, hence it can be used in high frequency devices. The dielectric property is strongly affected by the copper dopant in MnS nanocrystals. Impedance data has been analyzed using an equivalent circuit, enabling determination of grain and grain boundary contributions. The bulk resistance decreases with rise in temperature and exhibit a NTCR behavior. The electrical modulus has confirmed the presence of a hopping mechanism in the materials.

The frequency dependence AC conductivity at different temperature indicated that the conduction process is thermally activated process. The activation energy at 373K is found to be 0.25 eV and 0.17 eV for undoped and Cu-doped MnS nanocrystals respectively. The measured magnetization curve reveals that the MnS sample has clear paramagnetic behavior whereas superparamagnetic behavior appears on the Cu-doped MnS nanocrystals.

References

1. S.H. Wei, A. Zunger, Phys. Rev. B 48 (1993) 6111-6115.
2. Changhua Ana, Kaibin Tanga, Xianming Liub, Fanqing Lib, Guien Zhoub, Yitai Qiana, Journal of Crystal Growth 252 (2003) 575-580.
3. Ning Zhang, Ran Yi, ZhongWang, Rongrong Shi, HaidongWang, Guanzhou Qiu, Xiaohe Liu, Materials Chemistry and Physics 111 (2008) 13-16.
4. Juan Beltran-Huarac, Javier Palomino, Oscar Resto, Jingzhou Wang, Wojciech M. Jadwisienczak, Brad R. Weiner, Gerardo Morella, RSC Adv. 4 (2014) 38103-38110.
5. S. S. Aplesnin, L. I. Ryabinkina, O. B. Romanova, D. A. Balaev, O. F. Demidenko, K. I. Yanushkevich, N. S. Miroshnichenko, Phys. Solid State 49 (2007) 2080-2085.
6. Alaric D. Sangma, P. K. Kalita, J. Mater. Sci. Technol. 2 (2012) 57-60.
7. Reddy KR, Park W, Sin BC, Noh J, Lee Y, J Colloid Interface Sci. 335 (2009) 34-39.
8. Subhajit Biswas, Soumitra Kar, Subhadra Chaudhuri, Journal of Crystal Growth 299 (2007) 94-102.
9. Meiying Liu, Nannan Shan, Linlin Chen, Xiaoqian Li, Bona Li, Wansheng you, Applied surface science 258 (2012) 7922-7927.
10. YongCai Zhang, Hao Wang, Bo Wang, HaiYan Xu, Hui Yan, Masahiro Yoshimura, Optical Materials 23 (2003) 433-437.
11. A. Khorsand Zak, W.H. Abd. Majid, M.E. Abrishami, Ramin Yousefi, Solid State Sciences 13 (2011) 251-256.
12. Salah Abdul-Jabbar Jassim, Abubaker A. Rashid Ali Zumaila, Gassan Abdella Ali Al Waly, Results in Physics 3 (2013) 173-178.
13. B. Anderson, J. Gjonnes, A. R. Forouhi, J. Less-Comm. Metal. 61 (1978) 273-291.
14. Wei-Yu Wu, J. N. Schulman, Hsu, Uzi Efron, Appl. Phys. Lett. 51 (1987) 710-712.
15. R Hepzi Pramila Devamani, M Alagar, Nano Biomed. Eng. 5 (2013) 116-120.
16. Louse Barry, Mark Copley, Justin D. Holmes, David J. Otway, Olga Kazakova, Michael A. Morris, Journal of Solid State Chemistry 180 (2007) 3443-3449.
17. M. Y. Hao, S. Y. Lou, S. M. Zhou, R. J. Yuan, G. Y. Zhu, N. Li, Nanoscale Research Letters 7 (2012) 1-13.
18. N.Kandasamy, S.Saravanan, Deepak Ranjan Nayak, International Journal of Advanced Chemical Science and Applications 1 (2014) 25-30.
19. Rajeev Kayestha, Sumati, Krishnan Hajela, FEBS Letters 368 (1995) 285-288.
20. Yange Zhang, Zude Zhang, Shutao Wang, Xuchu Ma, Yitai Qian, Materials Chemistry and Physics 97 (2006) 365-370.
21. Yange Zhang, Zude Zhang, Shutao Wang, Xuchu Ma, Yitai Qian, Materials Chemistry and Physics 97 (2006) 365-370.
22. T. Veeramanikandasamy, K. Rajendran, K. Sambath, J Mater Sci: Mater Electron 25 (2014) 3383-3388.
23. Rajeev Kayestha, Sumati, Krishnan Hajela, FEBS Letters 368 (1995) 285-288.
24. C.M. Muiva, T.S. Sathiaraj, K. Maabong, Ceramics International 37 (2011) 555-560.

25. Zhijun Wang, Feng Tao, Feng Pan, Yufeng Sun, Weili Cai, Lianzeng Yao, Applied Surface Science 258 (2011) 44-49.
26. A.I. Borhan, A.R. Iordan, M.N. Palamaru, Mater. Res. Bull. 48 (2013) 2549-2556.
27. M. Atif, M. Nadeem, R. Grössinger, R. Sato Turtelli, Journal of Alloys and Compounds 509 (2011) 5720-5724.
28. Ahmet Altındal, Mustafa Coskun, Ozer Bekaroglu, Synth. Met. 162 (2012) 477-482.
29. Ji Young Kim, Min-Wook Oh, Seunghun Lee, Yong Chan Cho, Jang-Hee Yoon, Geun Woo Lee, Chae-Ryong Cho, Chul Hong Park, Se-Young Jeong, Scientific Reports, 4 (2014) 5450.
30. Subhanarayan Sahoo, Umasankar Dash, S. K. S. Parashar, S. M. ALI, Journal of Advanced Ceramics 2 (2013) 291-300.
31. M. Haj Lakhdar, B. Ouni, M. Amlouk, Materials Science in Semiconductor Processing 19 (2014) 32-39.
32. Dhananjay K. Sharma, Nawnit Kumar, Seema Sharma, Radheshyam Rai, Materials Chemistry and Physics 141 (2013) 145-152.
33. P. B. Macedo, C. T. Moynihan, R. Bose, Phys. Chem. Glasses 13 (1972) 171-179.
34. C. A. Angell, Chem. Rev 90 (1990) 523-542.
35. I. M. Hodge, M. D. Ingram, A. R. West, J. Electroanal. Chem 74 (1976) 125-143.
36. J.L. Izquierdo, A. Forero, G. Bolanos, V.H. Zapata, O. Moran, Solid State Sciences 38 (2014) 62-68.
37. Gagan Kumar, Sucheta Sharma, R.K. Kotnala, Jyoti Shah, Sagar E. Shirsath, Khalid M. Batoo, M. Singh, Journal of Molecular Structure 1051 (2013) 336-344.
38. Xiufeng Song, Renli Fu, Hong He, Microelectronic Engineering 86 (2009) 2217-2221.
39. F. Yakuphanoglua, Y. Aydogdua, U. Schatzschneiderb, E. Rentschlerb, Solid State Communications 128 (2003) 63-67.
40. E. Sarantopoulou, J. Kovac, S. Pispas, S. Kobe, Z. Kollia, A.C. Cefalas, Superlattices and Microstructures 44 (2008) 457-467.
41. Ying Wang, T.L. Tan, K.S. Tan, P. Lee, Hai Liu, Boluo Yadian, Ge Hu, Yizhong Huang, R.V. Ramanujan, R.S. Rawat, Applied Surface Science 315 (2014) 37-44.
42. Carolyn I. Pearce, Richard A.D. Pattrick, David J. Vaughan, Reviews in Mineralogy and Geochemistry 61 (2006) 127-180.
43. Remy Tappero, Albert Lichanot, Chemical Physics 236 (1998) 97-105.
44. S. Shafiu, R. Topkaya, A. Baykal, M.S. Toprak, Materials Research Bulletin 48 (2013) 4066-4071.
45. Yu. P. Kalmykov, W. T. Coffey, S. V. Titov, Physics of the Solid State 47 (2005) 272-280.

Chapter VII
SUMMARY

Semiconductor nanocrystals have attracted great fundamental and technical interest because of their unique optical and electrical properties. The present work deals with the synthesis and property characterizations of MnS nanocrystals and on their doping with Cu^{2+} ions. MnS and Cu-doped nanocrystals have been prepared by a simple and low cost wet chemical technique. The different synthesis strategies of MnS nanocrystals and their structural, optical, electrical and magnetic properties as well as some of the possible applications have been explored. The different synthesis conditions, especially the refluxing temperature of the growth solution and the precursor molar concentration have been used for this analysis. The optimization of the growth conditions and realizing the size controllable growth has been investigated. The characterization of as-synthesized MnS nanocrystals have been carried by using X-ray diffraction, Scanning electron microscopy, Transmission electron microscopy, Fourier transform infrared spectroscopy, UV-Visible spectroscopy, Photoluminescence spectroscopy, Impedance Spectroscopy and Vibrating sample magnetometer.

The aim of the performed work was threefold. The first aim was to study the influence of refluxing temperature on the structural, optical, electrical and magnetic properties of chemically synthesized MnS nanocrystals. The second aim was to study the influence of Mn/S molar ratio on the structural, optical, electrical and magnetic properties of chemically synthesized MnS nanocrystals. The third and final aim was to investigate the influence of Cu-doping on structural, optical, electrical and magnetic properties of chemically synthesized MnS nanocrystals.

XRD pattern revealed the existence of two phases. The metastable sphalerite phase β-MnS nanocrystals are formed in the sample with 55 °C refluxing temperature and 1:1 molar ratio. The β-MnS sphalerite to metastable γ-MnS wurtzite phase transition have been identified when the refluxing temperature is increased from 55 °C to 65 °C and when the molar concentration of manganese to sulfide is either 1:2 or 2:1. The XRD pattern of Cu-doped MnS nanocrystals showed the γ-wurtzite phase. The nanocrystalline wurtzite is stable at low temperature. The optimized reaction parameters for preparing wurtzite γ-MnS nanocrystals were the refluxing temperature higher than 65°C and molar ratio of manganese to sulfide is either 1:2 or 2:1. From the Williamson-Hall (W-H) Analysis, it is seen that crystallite size increases and the strain in the crystallites decreases with increasing refluxing temperature and value of crystallite size decrease with increasing copper dopant concentration. The crystallite size calculated from modified Scherrer formula was found to be decreased with increasing Mn concentration from 2.04 nm, on the other hand increased with increasing sulfide concentration.

The SEM analyses of MnS nanocrystals exhibit narrow particle size distribution and are spherical in shape. The morphology of the MnS nanocrystals changed from spherical into dendrites upon increase in the refluxing temperature above 75°C in the precursor solution.

It was also found that the processing temperature strongly affected the particle morphology. Also it was noted that particles get agglomerated at 2:1 molar concentration due to lack of effective capping and sulfur contents. TEM image shows that particles are nearly in spherical shape and having average particle size of approximately 175 nm to 300 nm for the pure MnS and 15 nm for cu-doped MnS nanocrystals.

FTIR spectroscopy revealed the absorption bands at 415 to 425cm^{-1} and 590cm^{-1} are due to the vibration of the chemical Mn-S bond and Mn-S-Cu bond, respectively. All the as-synthesized MnS and Cu doped MnS nanocrystals exhibit a blue shift when compared to that of the bulk MnS. An interesting fact is that the optical band gap correlates linearly with the precursor molar ratio, refluxing temperature and cu-doping which allows the tuning of optical bandgap. The optical bandgap tends to decrease with an increase of refluxing temperature and increase with an increase in Mn^{2+} ion concentration as well as Cu^{2+} ion concentration. The 4f and 5d states lead to strong absorption bands in the spectral range of 220 nm–300 nm. Hence, this material has potential application in UV detection and solar cells. PL spectra of as-prepared MnS nanocrystals revealed that the two peaks at around 415 nm and 440 nm are attributed to the violet-blue emissions. It has been observed that PL intensities decreasing with increasing refluxing temperature. Quantum confinement effect and its influence on absorption and photoluminescence spectra of MnS and Cu-doped MnS nanocrystals are studied. The emission peaks may be attributed to the recombination of charge carriers in deep traps of surface localized states and a photo generated hole caused by surface defects. The wide band gap of the MnS nanocrystals with blue emission makes it suitable for use in blue light emitting devices and opto-electronic devices.

Electrical measurements were performed in the frequency range of 50Hz-5MHz with 323K to 423K. The maximum dielectric constant decreases with the increase of frequency, which is characteristic of a relaxation phenomenon. The dielectric loss decreases with increase of frequency. It could be due to increase of space charge polarization. The AC conductivity increases and activation energy decreases with increase of frequency. It was confirmed that the conduction is temperature dependent and thermally activated process which is existing in the material. It was found that a non-Debye type of relaxation behavior in the material. The relaxational frequencies shifted to higher frequencies with the increase of temperature. The impedance data were analyzed in order to obtain the bulk resistance and grain boundary resistance. The grain and grain boundary resistance decreases with an increase in temperature which shows the negative temperature coefficient of resistance (NTCR) of the material. It is the behavior of a typical semiconductor. The dielectric, impedance, AC conductivity and activation energy of electrical studies in nanocrystalline MnS and cu-doped MnS exhibit clear size dependence.

It is observed that all the MnS samples of all the series exhibit a clear paramagnetic behavior whereas superparamagnetic behavior is observed on the Cu-doped MnS nanocrystals. It is shown that the magnetization and coercivity can be adjusted by Cu-doping in the MnS nanocrystals for possible sensors and electromagnetic resonance applications.

www.ingramcontent.com/pod-product-compliance
Lightning Source LLC
Chambersburg PA
CBHW060832220526
45466CB00003B/1072